The Water, Food, Energy and Climate Nexus

Global trends of population growth, rising living standards and the rapidly increasing urbanization of the world are intensifying the demand for water, food and energy. Added to this is the growing threat of climate change, which will have huge impacts on water and food availability. It is increasingly clear that there is no place in an interlinked world for isolated solutions aimed at just one sector. In recent years the "Nexus" has emerged as a powerful concept to capture these inter-linkages of resources and is now a key feature of policy-making.

This book is one of the first to provide a broad overview of both the science behind the Nexus and the implications for policies and sustainable development. It brings together contributions by leading intergovernmental and governmental officials, industry, scientists and other stakeholder thinkers who are working to develop the approaches to the Nexus of water–food–energy and climate. It represents a major synthesis and state-of-the-art assessment of the Nexus by major players, in light of the adoption by the United Nations of the new Sustainable Development Goals and Targets in 2015.

Felix Dodds is a Senior Affiliate of the Water Institute and a Senior Fellow at the Global Research Institute at the University of North Carolina, Chapel Hill, USA. He is also an Associate Fellow at the Tellus Institute and was the Executive Director of the Stakeholder Forum for a Sustainable Future from 1992 to 2012. He is author or editor of several books on sustainable development and resource security.

Jamie Bartram is a Don and Jennifer Holzworth Distinguished Professor of Environmental Sciences and Engineering and Director of the Water Institute at the University of North Carolina at Chapel Hill, USA. He is the author or editor of numerous academic papers and books, including the *Routledge Handbook of Water and Health* (2015).

Earthscan Studies in Natural Resource Management

The Water, Food, Energy and Climate Nexus
Challenges and an agenda for action
Edited by Felix Dodds and Jamie Bartram

Mitigating Land Degradation and Improving Livelihoods
An integrated watershed approach
Edited by Feras Ziadat and Wondimu Bayu

Adaptive Cross-scalar Governance of Natural Resources
Edited by Grenville Barnes and Brian Child

The Water, Energy and Food Security Nexus
Lessons from India for development
Edited by M. Dinesh Kumar, Nitin Bassi, A. Narayanamoorthy and M.V.K. Sivamohan

Gender Research in Natural Resource Management
Building capacities in the Middle East and North Africa
Edited by Malika Abdelali-Martini, Aden Aw-Hassan

Contested Forms of Governance in Marine Protected Areas
A study of co-management and adaptive co-management
Natalie Bown, Tim S. Gray and Selina M. Stead

Adaptive Collaborative Approaches in Natural Resource Management
Rethinking participation, learning and innovation
Edited by Hemant R. Ojha, Andy Hall and Rasheed Sulaiman V

Integrated Natural Resource Management in the Highlands of Eastern Africa
From concept to practice
Edited by Laura Anne German, Jeremias Mowo, Tilahun Amede and Kenneth Masuki

For more information on books in the Earthscan Studies in Natural Resource Management series, please visit the series page on the Routledge website: http://www.routledge.com/books/series/ECNRM/

The Water, Food, Energy and Climate Nexus

Challenges and an agenda for action

Edited by Felix Dodds and Jamie Bartram

First published 2016
by Routledge
2 Park Square, Milton Park, Abingdon, Oxon OX14 4RN

and by Routledge
711 Third Avenue, New York, NY 10017

Routledge is an imprint of the Taylor & Francis Group, an informa business

British Library Cataloguing-in-Publication Data
A catalogue record for this book is available from the British Library

Library of Congress Cataloging in Publication Data
Names: Dodds, Felix, editor. | Bartram, Jamie, editor.
Title: The water, energy, food and climate nexus : challenges and an agenda for action / edited by Felix Dodds and Jamie Bartram.
Description: London ; New York : Routledge, 2016. | Series: Earthscan studies in natural resource management | Includes bibliographical references and index.
Identifiers: LCCN 2015045224| ISBN 9781138190948 (hbk) | ISBN 9781138190955 (pbk) | ISBN 9781315640716 (ebk)
Subjects: LCSH: Sustainable development. | Water security. | Food security. | Energy security. | Climatic changes.
Classification: LCC HD75.6 .W376 2016 | DDC 333.7--dc23
LC record available at http://lccn.loc.gov/2015045224

ISBN: 978-1-138-19094-8 (hbk)
ISBN: 978-1-138-19095-5 (pbk)
ISBN: 978-1-315-64071-6 (ebk)

Typeset in Bembo
by Saxon Graphics Ltd, Derby

Contents

Acronyms and Abbreviations

BOLT	Build Operate Lease Transfer
CITES	Convention on International Trade in Endangered Species
CLEW	Climate, land, energy, water
CMS	Conservation of Migratory Species
CSD	Commission on Sustainable Development
DAC	Development Assistance Committee
EEZ	Exclusive Economic Zone
EU	European Union
FAO	Food and Agriculture Organization (UN)
FE2W	Food, Energy, Environment and Water Network
FFD	Financing for Development
GDP	Gross National Product
GEF	Global Environmental Facility (UN)
GHG	Greenhouse Gases
GRI	Global Research Institute
GWP	Global Water Partnership
ICLEI	International Council for Local Environmental Initiatives
ICSU	International Council for Science
IDP	Integrated Development Planning
IFAD	International Fund for Agricultural Development
IPCC	Intergovernmental Panel on Climate Change
IRENA	International Renewable Energy Agency
IWMI	International Water Management Institute
JPoI	Johannesburg Plan of Implementation
LA21	Local Agenda 21
LDC	Least Developed Countries
LTG	Limits to Growth
MDBs	Multilateral Development Banks
MDGs	Millennium Development Goals
NGOs	Nongovernmental Organizations
NIC	National Intelligence Council (United States)
OECD	Organization for Economic Cooperation and Development
OTEC	Ocean Thermal Energy Conversion

OWG	Open Working Group
PARM	Platform for Agricultural Risk Management
PPP	Public–Private Partnerships
PRI	Principles for Responsible Investment
PRS	Pearl Rating System
REEEP	Renewable Energy and Energy Efficiency Partnership
ROAD	Risk and Opportunities Assessment for Decision-making
SARD	Sustainable Agriculture and Rural Development
SDGs	Sustainable Development Goals
SEEA	System for Environmental Economic Accounting
SEI	Stockholm Environment Institute
SIDS	Small Island Developing States
SIWI	Stockholm International Water Institute
SWH	Solar Water Heaters
TEEB	The Economics of Ecosystems and Biodiversity
TFWW	The Future We Want
TNC	Commission on Transnational Corporations (UN)
UN	United Nations
UN CEB	UN Chief Executive Board
UN CSD	United Nations Commission on Sustainable Development
UNC	University of North Carolina at Chapel Hill
UNCED	United Nations Conference on Environment and Development
UNCFCCC	United Nations Framework Convention on Climate Change
UNDESA	United Nations Department of Economic and Social Affairs
UNDP	United Nations Development Programme
UNEP	United Nations Environment Programme
UNESCO	United Nations Educational, Scientific and Cultural Organization
UNGA	United Nations General Assembly
UNGC	United Nations Global Compact
WACC	Weighted Average Cost of Capital
WAVES	Wealth Accounting and the Valuation of Ecosystem Services
WBCSD	World Business Council for Sustainable Development
WEF	Water–Energy–Food Security Nexus
WEF	World Economic Forum
WHO	World Health Organization (UN)
WMO	World Meteorological Organization (UN)
WRI	World Resources Institute
WSSD	World Summit on Sustainable Development
WTO	World Trade Organization
WWF	World Wide Fund for Nature

Notes on contributors

Tony Allan specializes in the analysis of water resources in semi-arid regions and on the role of global systems in ameliorating local and regional water deficits. He pointed out that water-short economies achieve water and food security mainly by importing water-intensive food commodities, and coined the concept "virtual water." He is currently working on why the accounting systems in our food supply chain are dangerously blind to the costs of water and of misallocating it. In 2008 he was awarded the Stockholm Water Prize in recognition of his contribution to water science and water policy. In 2011 he became International Academic Correspondent of the Academy of Sciences of Spain. In 2013 he received the international Environmentalist Award from the Florence-based Bardini and Peyron Foundation, and the Monaco Water Award from Prince Albert II of Monaco.

Jamie Bartram is a Don and Jennifer Holzworth Distinguished Professor of Environmental Sciences and Engineering, and has been director of the Water Institute at the University of North Carolina at Chapel Hill since 2009. Bartram is author of more than ninety academic papers, author or editor of forty-nine books and author of more than fifty book chapters. From 1998 to 2009 he was based at the World Health Organization's headquarters. For most of this time he was the Coordinator of the Water, Sanitation, Hygiene and Health Unit which received international recognition for leadership in evidence-based policy and good practice. From 2004 to 2006 he also served as the first chair of UN Water.

David Le Blanc is an economist currently working in the Division for Sustainable Development in the United Nations Secretariat in New York. While at the UN he has worked on financing, integrated strategies for sustainable development, mining, agriculture and food, and sustainable development assessments. Prior to joining the UN he worked at the World Bank as a senior economist in the urban and housing finance groups.

Joachim von Braun is Director of the Center for Development Research (ZFF), Bonn University, and a professor of economic and technological change. His main research interests are in sustainable economic development, poverty reduction, food and nutrition security, resource economics, trade, science and technology policy. He is Chair of the Bio-economy Council of the Federal German Government, Vice-President of the NGO "Welthungerhilfe", Vice Chair of the Board of Global Alliance for Improved Nutrition (GAIN). He was Director General of the International Food Policy Research Institute (IFPRI) (2002–2008), and President of the International Association of Agricultural Economists (IAAE).

Kathrine Brekke is the Urban Research Officer in charge of maintaining and developing the research management activities at the ICLEI World Secretariat. She has a background in economics and development studies and Mastered in urban governance. She has coordinated various international research projects and a case study series on renewable energy policies in cities, and coordinated a study including two pilot projects on the Urban Nexus approach.

Jeb Brugmann is a strategist and innovation process expert for business, government and the nonprofit sector. He has been internationally recognized for his contributions to the establishment of the fields of urban sustainability planning, local climate action planning and "base of the pyramid" business development. He has worked with municipalities in twenty-eight countries and with a range of corporate clients to design award-winning products, new business ventures, programmes and policies. In 1989–1990 Jeb founded ICLEI-Local Governments for Sustainability, the largest international association of local governments. He served as ICLEI's Secretary General from 1991 to 2000.

Albert Butare is a former Minister of State for Infrastructure in the Republic of Rwanda and has over twenty years of energy, water and communication experience in Africa. He is skilled in high-level public sector policy and project development and implementation, engineering, social and economic development. He was a co-chair of the Bonn-led Nexus process all the way through to Rio. He is currently a CEO of Africa Energy Services Group Ltd based in Kigali, Rwanda with a branch in Dar es salaam (www.africaesg.com).

Paula Caballero Gomez is a senior director for the Environment at the World Bank. Prior to joining the World Bank in July 2014 she was the director for Economic, Social and Environmental Affairs in the Ministry of Foreign Affairs of Colombia. She was a leading voice and negotiator in international fora, including the UN Framework Convention on Climate Change, Rio+20 and the post-2015 framework. She was awarded a Zayed International Prize for the Environment in 2014 for her contribution to "environmental action leading to positive change."

Felix Dodds is a Senior Affiliate of the Water Institute, a Senior Fellow at the Global Research Institute at the University of North Carolina and an

Associate Fellow at the Tellus Institute. He was the co-director of the 2014 Nexus Conference on Water, Food, Energy and Climate. He was the Executive Director of Stakeholder Forum for a Sustainable Future from 1992 to 2012. He has written or edited twelve books. Prior to Rio+20 with Michael Strauss and Maurice Strong he co-wrote "Only One Earth: The Long Road via Rio to Sustainable Development." In 2011 he chaired the United Nations DPI 64th NGO conference – "Sustainable Societies Responsive Citizens." From 1997 to 2001 he co-chaired the UN Commission on Sustainable Development NGO Steering Committee. He is also an International Ambassador for the City of Bonn.

Pay Drechsel has twenty-five years of working experience as an environmental scientist aiming at the safe recovery of irrigation water, nutrients, and carbon from domestic waste streams, integrated natural resources management, and sustainable agricultural production in developing countries. Pay worked extensively on urban and peri-urban agriculture, safe wastewater irrigation in low-income countries, as well as economic opportunities for resource recovery and reuse. Pay joined the International Water Management Institute (IWMI) and for the last ten years has been coordinating different IWMI research programmes and divisions in the agriculture–sanitation interface, supporting impact-oriented and multi-disciplinary teams.

R. Quentin Grafton, FASSA is a Professor of Economics, the holder of the UNESCO Chair in Water Economics and Transboundary Water Governance, Executive Director of the Asia and Pacific Policy Society, and Director of the Centre for Water Economics, Environment and Policy (CWEEP) at the Crawford School of Public Policy at the Australian National University. He also serves as the Director of the Food, Energy, Environment and Water (FE2W) Network.

Ania Grobicki is Deputy Secretary General of the Ramsar Convention on Wetlands, where she advises on global policy, developmental, technical and scientific issues, and oversees the regional implementation of the Convention. Previously she was the Executive Secretary of the Global Water Partnership (2009 to 2015) where she managed the worldwide multi-stakeholder network and its global secretariat.

Munir A. Hanjra is currently working as an economist at the International Water Management Institute (IWMI). Previously he worked as a senior research fellow at the Charles Sturt University Institute for Land, Water and Society, and CSIRO Land and Water, Australia. He is a development economist with over twenty years of professional experience on issues related to sustainable water management.

Gary Lawrence was Vice President and Chief Sustainability Officer for AECOM Technology Corporation, a $20-billion global provider of professional technical and management support services. Gary leads

AECOM's sustainability efforts by managing its extensive resources and skills in sustainability for projects across the enterprise. In addition he is a Senior Fellow at the Design Futures Council, a Senior Fellow at the EastWest Institute for Water, Food, Energy, Security Nexus programme, and he is on a number of advisory boards related to sustainability and resilience.

Sylvia Lee has fifteen years of experience in the water sector and currently leads the water programme at the Skoll Global Threats Fund. Prior to that she worked at the World Bank based in Kathmandu, Nepal where she focused on transboundary water issues and climate change adaptation and resilience building. Lee also led the Water Initiative at the World Economic Forum based in Geneva, Switzerland, where her work focused on raising awareness of the global water challenge and engaging the private sector on water issues.

Iain MacGillivray is a Special Advisor to the President of the International Fund for Agricultural Development (IFAD) focusing on the Post-2015 Sustainable Development framework and on embedding nutrition throughout IFAD. His career spans farm management in Argentina, farming in the Sudan, integrated rural development in the Americas and marketing and consultative work for the private and public sectors. He is a strong supporter of food-based approaches to micronutrient malnutrition, and of bridging the divide between agriculture, nutrition and health. He was awarded CIDA's Presidential Award for Excellence for his work in developing Canada's Food Security Strategy.

Anna Delgado Martin is currently working as a consultant for the World Bank, providing strategic and technical support to the Thirsty Energy initiative in Morocco, South Africa, and China. Through her experiences, Delgado Martin has been gaining expertise on the Water, Energy and Climate Nexus for the past six years. She is also interested in social entrepreneurship and innovation and she has been engaging in several projects to promote water and energy solutions for rural areas.

Nathanial Matthews is the Global Research Coordinator at the CGIAR Program on Water, Land and Ecosystems (WLE). In his role with WLE, Matthews manages teams and projects across WLE's four focal regions (West Africa, East Africa, the Ganges and the Greater Mekong) working with over 175 partners to develop scalable solutions for reducing poverty, improving food security and maintaining healthy ecosystems. A political and environmental scientist by training, Matthews has fourteen years of professional experience across business, education, research and consulting that spans over twenty countries. He is a Senior Visiting Fellow at King's College London, a Visiting Fellow at the University of East Anglia Water Security Centre and a Fellow of the Royal Geographic Society.

Alisher Mirzabaev is a senior researcher at the Center for Development Research (ZEF), University of Bonn. His current research areas include the economics of land degradation, bioenergy and the Water–Energy–Food Security Nexus. Before joining ZEF in 2009 he was an economist with the International Center for Agricultural Research in the Dry Areas (ICARDA). He holds a PhD from the University of Bonn in Germany, writing his thesis on the economics of climate change in Central Asia.

Divyam Nagpal is an Associate in IRENA's Knowledge, Policy and Finance Centre. Within the Policy Unit he works on a broad range of projects, including renewable energy policy assessment, off-grid renewables for energy access, and the Water, Energy and Food Nexus. He has co-authored several publications, including IRENA's recent publication on renewable energy in the Water, Energy and Food Nexus.

David Norman was Senior Manager for Sustainable Development Policy at SABMiller from 2013 to 2015, responsible for the company's engagement in policy processes across sustainable development issues. Before this role, he was WWF-UK's Director of Campaigns and Director of External Affairs, working on a wide range of environmental and development issues with UK and European policy-makers and businesses. He was previously Director of Policy and Communications for the UK's Liberal Democrat Party, co-chair of the Commonwealth Education Fund's Management Committee, and Global Education Adviser for Save the Children UK. David now runs Global Insight Consulting, helping companies, governments and NGOs to build stronger partnerships and strategies for sustainable development.

Stuart Orr has a background in the private sector and academic research and joined WWF in 2006. His work explores the role of the private sector with regard to development and specifically water-related issues. Orr has published mainly on water measurement, agricultural policy and water-related risk, with some specific work on food, water and energy in relation to biofuels and dam development. He also sits on a number of sustainability boards and initiatives.

Diego Rodriguez is currently a Senior Economist at the Water Global Practice of the World Bank Group. He is the Task Team Leader of Thirsty Energy, a World Bank initiative to assist countries to quantify the trade-offs of energy and water planning and investments. He is also the thematic leader on climate change, variability and uncertainty and is working on the use of decision scaling tools to identify climate risks and build resilience of water infrastructure investments. Prior to joining the World Bank he worked for fifteen years at the Inter-American Development Bank and also worked at the Danish Hydraulic Institute.

Cole Simons is a current junior at the University of North Carolina at Chapel Hill. He is a dual public policy and peace, war, and defense major. He has previously supported the production of two books, "Governance for

Sustainable Development: Ideas for the Post 2015 Agenda" and "The Plain Language Guide to Rio+20: Preparing for the New Development Agenda." He serves as the Oversight and Advocacy Chair of UNC's Student Congress where he is the primary advocate of student opinion to university administration and the state legislature. Simons plans to graduate in spring 2017 before pursuing a career in government.

Antonia Sohns is a Water and Energy Analyst working on the Thirsty Energy initiative at the World Bank. Sohns holds an MSc from the University of Oxford in Water Science, Policy and Management, and a BSc in Earth Systems, Oceans Track from Stanford University. She has worked for non-governmental organizations in Thailand and Washington, DC.

Liz Thompson was appointed in 2010 as Assistant Secretary General of the United Nations and Executive Coordinator for the UNCSD Rio+20. She is currently Senior Adviser to the UN Secretary-General's Sustainable Energy for ALL (SE4ALL) initiative. Her experience includes acting as Special Adviser to the President of the UN General Assembly. Liz was an elected parliamentarian and Senator in Barbados where she held a number of ministerial portfolios including Minister of Energy and Environment, and where she led the development of Barbados' national policies on sustainable development, sustainable energy and a green economy; this latter was the first in the Americas. Her contributions to policy in sustainable development and climate change were recognized when she was awarded UNEP's Champion of the Earth Award 2008.

Steve Waygood is the Chief Responsible Investment Officer, Aviva Investors and leads Aviva Investors' Global Responsible Investment team. This team is responsible for integrating environmental, social and corporate governance (ESG) issues across all asset classes and regions of the c. £250 billion of assets under management. Steve founded the Sustainable Stock Exchange initiative as well as the Corporate Sustainability Reporting Coalition, which is aiming to catalyze a UN Convention promoting enhanced corporate transparency and integrated reporting. In 2011 he received the Yale Rising Star in Corporate Governance Award, and he was among the Financial News Top 100 Rising Stars in 2009.

Frank Wouters possesses over twenty-five years of international experience in the field of energy. He held several senior management positions in leading organizations and institutions in the fields of energy and sustainability, including Evelop International BV in the Netherlands, Econcern in Germany, NICE International in The Gambia, and TDAU at the University of Zambia. He currently provides strategic support to companies and organizations developing sustainable energy businesses worldwide, with a focus on structuring and finance. Wouters has worked throughout his career with a wide variety of stakeholders, including the private sector and government officials at the highest levels. From 2009 to 2012 he served as the Director of Masdar Clean Energy, a developer and operator of renewable power generation projects, where he was responsible for projects representing enterprise value of more than $3 billion in Asia, Africa and Europe. He was appointed Deputy Director-General of the International Renewable Energy Agency (IRENA) in September 2012, a position he held for two years.

Nicholas You is a veteran urban specialist and thought leader. He is a former chairman of the UN-Habitat World Urban Campaign Steering Committee, the Assurance Group for Urban Infrastructure of the World Business Council for Sustainable Development, and is currently the Chair of the Urban Strategy and Innovation Council for GDF-Suez. He also serves as a member of the board of Citiscope, the Huairou Commission and the Joslyn Institute for Sustainable Communities, and he is a fellow of the Guangzhou Institute for Urban Innovation and the Centre for Livable Cities, Singapore.

Foreword

From Bonn 2011 Nexus to Chapel Hill via Rio+20

Albert Butare

Genesis; the Bonn 2011 Nexus Conference

The basis of the Bonn 2011 Nexus Conference was anchored on the water, energy and food resource use and management practices prevailing today. The analysis in this context revealed that although significant progress has been made in improving people's standards of living, secure climate change imperatives and the associated ingredients including water, energy and food supplies still remain far from being achieved globally. Basic services are not available to the "bottom billion" – about 1.1 billion are without adequate access to safe water, close to 1 billion are undernourished and 1.5 billion are without access to modern forms of energy. People thereby remain deprived of their human rights and are trapped in poverty. For many others outside the bottom billion who have access to a basic level of service, the system does not yet offer the resources needed to raise their livelihoods and allow them to emerge from poverty.

With 70 percent of the global population of about 9.6 billion people projected to be living in cities by 2050, demands for water, energy and food will increase exponentially. With forecasts of a 70 percent increase in agricultural demand by 2050, energy demand increase of 40 percent by 2030 and water demand projections to satisfy agriculture and energy production in a similar order of magnitude, threats prevail.

Looking at this situation closely, pressures on natural resources have grown to such an extent over the past fifty years that they are challenging the effectiveness of conventional planning and decision-making. Trying to meet demand through single sector approaches, in response to what are inherently interlinked and interdependent processes, is limiting our ability to provide basic water, food and energy services to the poorest in the world, as well as failing to cope with the climate change that is continuously becoming unresponsive to us. Concurrently, there are signs of constrained growth, raised social and geopolitical tensions and irreversible environmental damage.

Failing to recognize the consequences of one sector on another, weak governance systems, limited awareness, distortions from perverse subsidies and lack of commitment can exacerbate a set of unintended consequences from decisions in one area on to another. The result is the sub-optimal allocation of resources and an inability to meet accepted norms of water, energy and food. We are reaching, and in some cases have already exceeded, the limit of resource availability and therefore need to build more innovative and sustainable solutions.

With this understanding, the Bonn Conference attempted to present initial evidence on how a Nexus approach can enhance water, energy and food security by increasing efficiency, reducing trade-offs, building synergies and improving governance across sectors. It also tried to underpin policy recommendations, which are detailed in a separate paper.

The key Nexus messages that came out of Bonn were that:

- A Nexus perspective increases the understanding of the interdependencies across water, energy, food and other policies such as climate and biodiversity;
- Understanding the Nexus is needed to develop policies, strategies and investments to exploit synergies and mitigate trade-offs among these development goals with active participation by and among government agencies, the private sector, academia and civil society;
- It was thus found important to incorporate the Nexus perspective in Rio+20 as well as in local, national and other international planning activities focusing on either water, food, energy or climate.

Nexus Opportunities: applying a Nexus approach was found to be an urgent need to identify the policy levers to implement a common future agenda and to create new opportunities for achieving these goals, while simultaneously reducing tensions between sectoral objectives. Realizing these opportunities requires action for change in key fields:

1. Increase policy coherence by ensuring that synergies and trade-offs among water, energy and food are identified in both the design and implementation of policies, plans and investments;
2. Accelerate access by progressively realizing – in a more coordinated way – the human rights obligations related to water, sanitation, energy, food and climate change to reap the resulting health, productivity and development benefits;
3. Create more with less by increasing resource productivity, establishing mechanisms to identify the optimal allocation of scarce resources for productive purposes;
4. End waste and minimize losses by reducing waste and losses along supply chains to capture significant economic and environmental gains and by changing mindsets and incentivizing technological development to turn waste into a resource and manage it for multiple uses;
5. Value natural infrastructure by investing to secure, improve and restore the considerable multi-functional value of biodiversity and ecosystems to

provide food and energy, conserve water, sustain livelihoods and contribute to a green economy;
6 Mobilize consumer influence by acknowledging and actively utilizing the catalyzing role that individuals have in choosing consumption patterns on such resources.

From Bonn to Rio

The UN Conference on Sustainable Development (Rio+20) was meant to address two key themes among other important and relevant aspects: the green economy within the context of sustainable development and poverty eradication, and the institutional framework for sustainable development. As a specific contribution to the Rio+20 process, the Bonn initiative intended to position the Water, Energy and Food Security Nexus in Green Economy concepts.

Bonn 2011 did provide a first platform for consideration of the close interlinkages of water, energy and food security and the benefits of a Nexus perspective in a multi-stakeholder process. The specific message from Bonn to Rio was that the outcome of Rio+20 should take into account and address the interdependencies between water, energy and food and act upon the challenge to make the Nexus work for all of us, but especially the poor. The Nexus approach is very much at the heart of the overall challenge of transforming our economies to green economies by changing growth patterns to become more sustainable. Making this happen will require us to identify the forces that are driving the adoption of a Nexus approach and to build alliances with them; this includes supporting leaders and champions at every level to take ownership and to develop the business case for integrated and sustainable solutions. Considering any change of approach is a particular challenge in today's economic climate, but the consequences of inaction will progressively limit our ability to deliver on our commitments and result in increasingly severe consequences for people's welfare, economic growth, jobs and the environment. The costs of inaction are too high. We need to act now.

From Rio to Chapel Hill

Coming from a part of the world that needs the Nexus approach the most, right from Bonn through to Rio and after, my preoccupation has been on how fast this Nexus philosophy should propagate and make an impact on the world in general and on the developing world in particular.

The Chapel Hill conference brought in a climate dimension to the original Bonn Nexus framework, and it did work very well. It was clearly missing as a conspicuous element in Bonn 2011, and I think Chapel Hill helped us to understand the Nexus in wider terms, both in its challenges and its opportunities. Even for those who question climate change, the realities of how the observed science (that governs climate) functions within the Water, Energy and Food Nexus cannot be denied.

The food sector alone contributes about a third of total greenhouse gas emissions through energy use, land use change, methane emissions from livestock and rice cultivation and nitrous oxide emissions from fertilized soils. At the same time, climate change mitigation places new demands on water and land resources and also impacts on biodiversity. Climate adaptation measures, such as intensified irrigation or additional water desalination, are often energy intensive. Thus climate policies can impact on water, energy and food, and adaptation action can in fact be maladaptive if it is not well aligned in a Nexus approach and implemented by appropriately interlinked institutions.

Since many of the original Steering Committee members for the Bonn Conference happened to be part of the Chapel Hill Advisory Committee or attendees of the Conference, we were able to create a unique opportunity to build on the Bonn Conference and gain an even greater understanding of the Nexus.

During the Chapel Hill conference we all had the chance to examine the connections and linkages between each of the four areas (Water, Food, Energy and Climate). The Bonn 2011 Conference aimed to provide input to the Rio+20 Summit, and the Chapel Hill 2014 Conference aimed to continue that journey towards our sustainable future by contributing to the Sustainable Development Goals process.

Foreword

Paula Caballero Gomez

There is broad consensus that the Millennium Development Goals (MDGs) fast-tracked action on some of the most pressing social challenges, and spearheaded action that otherwise would not have taken place. Yet there is also growing recognition that the development challenges of this century demand approaches that are unprecedented in terms of ambition and scale. The Sustainable Development Goals (SDGs) have the potential to be the platform for delivery.

The task for the coming decades is monumental. Human society has never been so interdependent, because there has never been so much potential for collective well-being, and because there has never been such a convergence of threats. We must continue to deliver long-term socio-economic gains for both growing and aging populations, which means not just tackling poverty but ensuring people do not fall back into poverty, reversing climate trajectories that currently point to a temperature rise of 3 to 4 degrees by 2100, and changing consumption and production patterns that are depleting and polluting Earth's systems at unmanageable rates – to name just a few of the challenges we collectively face. However, there is growing recognition of the cost and implications of business-as-usual trends.

As I write, the SDGs have just been formally adopted with acclaim by the international community, after what is probably the most inclusive process in human history. The degree of ownership is unprecedented. The private sector is actively engaged and showing decisive leadership on several targets. Many local authorities are working with the agenda to calibrate their development strategies. Civil society is ensuring that many voices continue to be heard and that strong accountability is built into the system. At country level, already a first "SDG Minister" has been named and many countries are quickly establishing inter-ministerial platforms to take on this new agenda.

However, the promise of the SDGs lies entirely on how they are implemented, and the core of this implementation is precisely the focus of this book: the need for integration across sectors, silos, scales and stakeholders – the Nexus approach. As communities, constituencies, countries and corporations grapple with translating the potential of the SDGs into action, this timely book will help decision-makers, scientists, practitioners, entrepreneurs and others identify core entry points for tapping into synergies and for explicitly addressing trade-offs and externalities.

In this undertaking, it will be fundamental that all stakeholders firmly embed in the Nexus approach the principles of sustainability and resilience, but also of inclusion. Application and implementation of the SDGs will be, as this publication highlights, strongly context-specific. In assessing and defining opportunities for action, it will be key that dedicated efforts are made to ensure that those that so often get left behind are explicitly targeted – in all countries. The inequalities that are becoming increasingly prevalent in all regions would undermine this universal agenda.

It will also be critical in taking Nexus approaches forward that global public goods are fully and explicitly integrated into policy and investment decisions. Work on planetary boundaries is what frames the new agenda. The urgency of the task ahead is, in many cases, driven by the irreversibility of trends triggered by man's footprint on the planet. Whether it be climate change, declining fish stocks and biodiversity losses, shifts in hydrological systems or land-use changes, the window of time for positive change is increasingly narrow. In this sense, too, this is a universal agenda.

The publication of this book, so soon after the adoption of the SDGs, is a cogent confirmation of the fact that many appreciate the urgency of moving forward. The core of effective SDG implementation is the Nexus approach, and this work should contribute to jump-starting the process.

CLARENCE HOUSE

Looking forward to the future and the enormous challenges embedded in the ambition to improve living conditions for the world's ever-expanding billions of people, I think it is fair to say that if we had a choice most of us would not start from here.

Economic growth has assuredly secured comfort and convenience for some, yet the scourge of poverty still afflicts a substantial proportion of humankind. Finding ways to achieve more growth will surely tax our ingenuity when we have already exceeded so many planetary boundaries, but it seems to me that we certainly won't be able to achieve this based on recent patterns of development.

This is not least because of how the natural systems that sustain human wellbeing are showing signs of serious stress. Climate change and its ever more obvious impacts are exacting an ever more costly human and economic toll; the disappearance of biodiversity and the degradation of ecological services are making us less secure and closing off options for the future. Yet the choices we make so often seek simple trade-offs, not least with regard to the long-term damage to ecological systems that we have come to regard as the inevitable price of short-term economic growth.

While there has been some progress in protecting the environment from the worst excesses of rapid development, what we have achieved to date is clearly insufficient, especially in the face of still rapidly expanding demand arising from unsustainable population and economic growth. Part of the problem comes with our habit of addressing the challenges that face us by taking action in separate silos, pursuing one subject at a time and, in so doing, often not seeing the fundamental connections between them.

When one examines the projections for increased demand for food, water and energy it becomes only too obvious that a more integrated approach is urgently needed. Crops need water and healthy soils in which to grow. Many of our most important energy sources also rely on copious quantities of water, including hydroelectric dams. With these dependencies in mind, we should perhaps not be too surprised to see the impact of drought in part reflected in higher food prices and, in some parts of the world, in power shortages.

The water that underpins our economy and food security is, of course, in turn dependent on functioning natural systems for its replenishment, including forests, wetlands and healthy soils. The more we degrade these, often through our relentless pursuit of industrialized agricultural production, the more we imperil the prospects for future development. All this leads me to conclude that it is not, first and foremost, facts and data about the changes taking place in our world of which we are most in need, but rather the leadership to adopt the kind of integrated approach that would enable us to move beyond our presently fragmented approach and to see that all the issues on the table are in fact related, not only to one another but also to the health of the natural world.

In the pages that follow, some of the world's leading thinkers and practitioners set out ideas on how best to approach this dense knot of connected questions, including the tight nexus of challenges that emerge in the interconnections between rising demand for food, energy and water.

Faced with such daunting challenges, it may be possible for us to navigate the choppy straits ahead, if only we can muster the will to harness new ideas, technologies and development strategies that will move us towards the sustainable and circular economy that we can increasingly see is within our grasp. For anyone seeking insight as to why and how we might do this, this book is a marvellous place to start.

Introduction

Felix Dodds and Jamie Bartram

In November 2011 the German government held a very ambitious global conference on the Nexus between Food, Water and Energy. It resulted in a growing recognition that what is needed is a movement away from a sector-by-sector approach to policy, science and practice, towards a more interlinked approach. The 2012 Rio+20 conference confirmed this need by agreeing that a new set of Sustainable Development Goals (SDGs) should be negotiated, and that they should reflect the much more complex world in which we now live. This book builds on these events and is a direct result of a second conference on the Water–Food–Energy–Climate Nexus, which was held in March 2014 in Chapel Hill, North Carolina. The book draws together some of the major thinking that has already begun to address some very challenging predictions that will frame the world we live in over the next fifteen years.

This is one of the first books to contribute some significant thinking and policy recommendations to the challenges for the new set of Sustainable Development Goals, which were adopted by Heads of State on September 25, 2015.

The implementation of these goals will require a much more interlinked approach than was implemented for the Millennium Development Goals (MDGs) in the previous fifteen years.

We have broken the book into a number of parts, which outline the policy and research agendas that will be needed to underpin the implementation of the Sustainable Development Goals on Water, Energy, Food and Climate.

Part 1: Learning from the past, building a new future: Nexus Scientific Research

Part 1 reviews how we have developed policy and research since the 1992 Earth Summit in the four Nexus areas. It then looks to the future and what new research is needed.

Part 2: Urban challenges of the Nexus: Local and global perspectives

Part 2 focuses on the urban area and its hinterland, demonstrating that this is where the Nexus needs managing. All communities need water, energy and food, and all will be impacted by climate change. What can be done in terms of planning, learning from good case studies and looking outside the traditional silo approaches?

Part 3: Natural resource security for people: Water, food and energy

Part 3 looks at the interlinkages from a water- and resource-based perspective. As we are still addressing many problems using a sector-based approach, what can be done by the water sector to address challenges in the other sectors?

Part 4: Nexus perspectives: Energy: Water and climate

Part 4 looks at the interlinkages from an energy perspective. As with water above, what can be done by the energy sector to address challenges in the other sectors?

Part 5: Nexus perspectives: Food, water, and climate

Part 5 looks at the interlinkages from an agriculture and food perspective. As with water and energy, what can be done by the agriculture and food sector to address challenges in the other sectors?

Part 6: Nexus corporate stewardship: How business is improving resource use

The final part looks at what the corporate sector can do in partnerships, developing norms and principles for addressing the Nexus and examining how the finance sector is starting to deal with the additional risks that the Nexus brings.

We have a number of people to thank for this book's development: David Edwards and Tony Juniper from the Prince Charles International Sustainability Unit (ISU); and the International Advisory Board for the 2014 Conference, which included many who have provided chapters for the book. In particular, we would also like to thank Cole Simons and Leah Komada, students of the University of North Carolina, for help with editing, and at Earthscan, Tim Hardwick and Ashley Wright.

We believe that the adoption of the Sustainable Development Goals offers a real opportunity for all of us to contribute to addressing the huge challenges ahead. The Nexus approach is an attempt to better understand the complex world we live in, and to help construct a research and policy agenda under the

SDGs that will help us along the path to a more sustainable, fair and equitable world. Through these challenging times the world needs a strong, robust and nimble United Nations, which we continue to believe has the credibility and capacity to be a force for global good and sustainability.

Many of the problems we will face on energy security, food security, water security and climate change will require multilateral solutions. The role of science in informing policy makers has never been more critical, as Maurice Strong, the UN Secretary-General at the Stockholm (1972) and Rio (1992) Conferences, eloquently puts it:

> Indeed, it has never been more important to heed the evidence of science that time is running out on our ability to manage successfully our impacts on the Earth's environmental, biodiversity, resource and life-support systems on which human life as we know it depends. We must rise above the lesser concerns that preempt our attention and respond to the reality that the future of human life on Earth depends on what we do, or fail to do in this generation. What we have come to accept as normal is not normal, as increased human numbers, the growing intensity of human impacts and the demographic dilemma faced by so many nations are returning the Earth to the conditions that have been normal for most of its existence that do not support human life as we know it.

Part 1

Learning from the past, building a new future:

Nexus scientific research

1 The history of the Nexus at the intergovernmental level

Felix Dodds and Jamie Bartram

Introduction and background

This chapter explores the development of water, energy and food as sectors and how they have come into conflict at the intergovernmental level in relation to the global sustainable development agenda set in Rio in 1992, followed by the development at the CSD, and revisited and developed further in New York at Rio+5 (1997) and in Johannesburg at Rio+10 (2002) and Rio+20 (2012). It looks especially at how, as Rio+20 approached, there was an increased realization that these three sectors needed to be viewed in an interlinked way – a Nexus perspective. In the post Rio+20 phase, within the Sustainable Development Goals (SDG) Open Working Group (OWG) strong voices promoted the idea that a Nexus perspective should be reflected in the targets and indicators under the relevant sector goals.

Fifteen years into the *new* millennium and well into the second century of ordered international relations under a single mechanism, i.e. the United Nations, it became clear that the predictions that were made in preparation for the first United Nations (UN) Conference on Human Environment in Stockholm in 1972 have largely come to pass. The stage for Stockholm was set by three ground-breaking reports. One of these was the Club of Rome's 'Limits to Growth' (LTG) that modelled and projected five variables (Figure 1.1): world population, industrialization, pollution, food production, and resource depletion.

Adopting a new approach to alignment, which does not immediately fit with the international policy machine, or indeed the machinery of the UN's member states which it largely reflects, is a significant risk that is not to be undertaken lightly. So why bother?

Among the future scenarios explored in Limits to Growth, scenario 1 reflects business as usual: the world proceeds in a traditional manner without any major deviation from the existing economic policies pursued during most of the twentieth century. Population and production increase until growth is halted by decreasing access to non-renewable resources. Increased investment is required to maintain resource flows. Finally, a lack of investment funds in the other sectors of the economy leads to declining output of both industrial goods

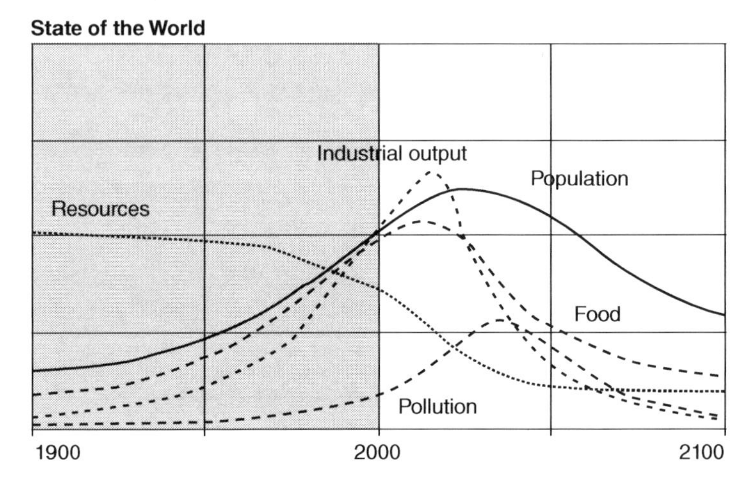

Figure 1.1 Limits to Growth: Business as Usual.

and services. Food and health services are reduced, decreasing life expectancy and raising average death rates (Meadows et al., 1972).

The report projected that the trends of all five variables lead rapidly to a point of environmental, economic and social collapse. Limits to Growth caused widespread controversy when it was published. Thirty-five years later, Graham Turner of the Commonwealth Scientific and Industrial Research Organization in Australia produced "A Comparison of 'The Limits to Growth' with Thirty Years of Reality" (2008) which examined the 1972 predictions of the Club of Rome with observed events. Chillingly, he found that predictions and observed events were consistent with the world's path through the early twenty-first century.

In 2014, Graham Turner concluded that:

> Regrettably, the alignment of data trends with the LTG dynamics indicates that the early stages of collapse could occur within a decade, or might even be underway. This suggests, from a rational risk-based perspective, that we have squandered the past decades, and that preparing for a collapsing global system could be even more important than trying to avoid collapse.
>
> (Turner, 2014)

Short history of how we got here

The 1972 Stockholm Conference reflected emerging Western concerns about the combined effects of an increasing global population, the finite or apparently finite nature of several critical resources, and early signals of humankind's ability to disrupt planetary systems through intent, pollution or resource use. One of the significant outcomes of the conference was an agreement to establish a new UN organ, the Environment Programme (UNEP). The UNEP would play a significant role as custodian of the environment, establishing UN conventions such as the Convention on International Trade in Endangered Species of Wild Fauna and Flora (CITES, 1973), the Convention on the Conservation of Migratory Species of Wild Animals (CMS, 1979), the Vienna Convention for the Protection of the Ozone Layer (1985), and the Regional Seas Conventions. These were some of the first environmental conventions. They did represent a substantive uptick in international attention to environmental concern.

At the ten-year review of the Stockholm Conference, the Canadian government suggested that a UN Commission on Environment and Development was required to balance the need for economic development with the protection of the environment. In 1983 the UN General Assembly established the World Commission on Environment and Development, popularly known as the Brundtland Commission after its Chair, the Norwegian Prime Minister Gro Harlem Brundtland. The Commission produced a report entitled 'Our Common Future' for the UN General Assembly in 1987, which would be the first UN document to define sustainable development as:

> development that meets the needs of the present without compromising the ability of future generations to meet their own needs.
>
> (UN, 1987)

This definition played to the growing concern about climate change. One of the most significant developments in the 1980s had been the establishment of the Intergovernmental Panel on Climate Change (IPCC) as a result of World Climate Conferences that had been held under the joint auspicious of the UNEP and the World Meteorological Organization (WMO). The first Assessment Report (1990) from the IPCC served as the core tenets of the UN Framework Convention on Climate Change (UNFCCC).

The Executive Summary of the Policymakers' Summary of the Working Group I Report reads:

> We are certain of the following: there is a natural greenhouse effect … emissions resulting from human activities are substantially increasing the atmospheric concentrations of the greenhouse gases: CO_2, methane, CFCs and nitrous oxide. These increases will enhance the greenhouse effect, resulting on average in an additional warming of the Earth's surface. The main greenhouse gas, water vapor, will increase in response to global warming and further enhance it.
>
> We calculate with confidence that: … CO_2 has been responsible for over half the enhanced greenhouse effect; long-lived gases would require immediate reductions in emissions from human activities of over 60 percent to stabilize their concentrations at today's levels …
>
> Under the IPCC business as usual emissions scenario, an average rate of global mean sea level rise of about 6 cm per decade over the next century (with an uncertainty range of 3–10 cm per decade), mainly due to thermal

> expansion of the oceans and the melting of some land ice. The predicted rise is about 20 cm ... by 2030, and 65 cm by the end of the next century.
>
> (IPCC, 1990)

The UN Conference on Environment and Development (UNCED), or Earth Summit, in 1992 was organized by Maurice Strong, who had also been the UN Secretary-General at the Stockholm Conference in 1972. This was probably the most important conference the UN has held on sustainable development.

The conference produced or incubated the most 'hard law' (legally binding agreements) of any UN conference. These were: the UN Framework Convention on Climate Change (1992); the UN Convention on Biological Diversity (1992); the UN Convention to Combat Desertification (1994); the Straddling Fish Stocks Agreement (1995); the Rotterdam Convention on the Prior Informed Consent Procedure for Certain Hazardous Chemicals and Pesticides in International Trade (1998); and the Stockholm Convention on Persistent Organic Pollutants (2001).

The Earth Summit also produced Agenda 21 – the UN Blueprint for the Twenty-first Century, the Rio Declaration – 27 Principles for Sustainable Development, forestry principles to guide more sustainable forestry, and the setup of the UN Commission on Sustainable Development which would monitor the implementation of these agreements.

Agenda 21 had forty chapters that dealt with environmental, social, economic and governance aspects of sustainable development. However, Agenda 21 did not contain a chapter on energy, since the United States and oil states had lobbied successfully for this not to be included. This critical omission would be reversed five years later when governments adopted a number of new chapters to Agenda 21 addressing transport, energy and tourism.

Importantly, Agenda 21 did include a chapter on water that itself built on the Dublin Principles that had been agreed the previous year. Through a preparatory process dedicated specifically to water, the principles have achieved enduring and widespread attention:

> Principle 1: Fresh water is a finite and vulnerable resource, essential to sustain life, development and the environment.
>
> Principle 2: Water development and management should be based on a participatory approach, involving users, planners and policy-makers at all levels.
>
> Principle 3: Women play a central part in the provision, management and safeguarding of water.
>
> Principle 4: Water has an economic value in all its competing uses and should be recognized as an economic good.
>
> (Dublin Principles, 1991)

The emphasis placed by the Dublin Principles on water having economic value was contested by non-governmental organizations (NGOs), who saw water primarily as a universal human right. This issue would be underlined by the World Water Forum in 2000 when the World Commission on Water for the 21st Century published its World Water Vision Commission Report, saying:

> Pricing of water services at full cost. Making water available at low cost, or for free, does not provide the right incentive to users. Water services need to be priced at full cost for all users, covering all costs related to operation and maintenance for all uses and investment costs for at least domestic and industrial uses. The basic water requirement needs to be affordable to all, however, and pricing water services does not mean that governments give up targeted, transparent subsidies to the poor.
>
> (World Water Commission, 2000)

This tension between water as an economic and as a social good would overshadow much of the water discussion until November 2002, when the UN Committee on Economic, Social and Cultural Rights adopted General Comment No. 15, formulated by experts and reading:

> The human right to water entitles everyone to sufficient, safe, acceptable, physically accessible and affordable water for personal and domestic uses. An adequate amount of safe water is necessary to prevent death from dehydration, to reduce the risk of water-related disease and to provide for consumption, cooking, personal and domestic hygienic requirements.
>
> (UN, 2002a)

General Comment 15 recognizes water not only as a limited natural resource and a public good, but also as a human right. The Comment's adoption was seen as a decisive step towards the recognition of water as a human right. However, it took until September 2010 and the Fifteenth Session of the UN Human Rights Council to pass a resolution reaffirming and refining an earlier General Assembly resolution (64/292). The refinements included the elimination of reference to international finance supports, which had prompted some, mainly industrialized nations to abstain from the earlier UNGA resolution.

> the right to safe and clean drinking water and sanitation as a human right that is essential for the full enjoyment of life and all human rights … the human right to safe drinking water and sanitation is derived from the right to an adequate standard of living and inextricably related to the right to the highest attainable standard of physical and mental health, as well as the right to life and human dignity.

In the area of Agriculture, Chapter 14 of Agenda 21 dealt with the idea of 'Promoting Sustainable Agriculture and Rural Development' (SARD) for the first time in a UN document. It described the food challenge in stark terms:

> By the year 2025, 83 per cent of the expected global population of 8.5 billion will be living in developing countries. Yet the capacity of available resources and technologies to satisfy the demands of this growing population for food and other agricultural commodities remains uncertain.
>
> (UN, 1992)

Agenda 21 saw the development of SARD as a mechanism to address this challenge.

The lost decade – drivers

> In recognition of the special importance of the role which can be fulfilled only by official development assistance, a major part of financial resource transfers to the developing countries should be provided in the form of official development assistance. Each economically advanced country will progressively increase its official development assistance to the developing countries and will exert its best efforts to reach a minimum net amount of 0.7 per cent of its gross national product at market prices by the middle of the Decade.
>
> (UN, 1970)

The 1992 Earth Summit happened against the backdrop of the recent dissolution of the Soviet Union bloc in 1990, and there was a strong belief in a 'peace dividend' that might fund the delivery of Agenda 21. The Secretary-General at the Summit, Maurice Strong, was asked how much it would cost to implement the blueprint for sustainable development. The resulting UN estimate was US$625 billion a year, with a transfer of US$125 billion from developed to developing countries. However, instead of the expected positive change, the 1992 aid flows of US$90 billion (0.32 percent of GNI) fell to 0.22 percent of GNI by the 1997 Earth Summit Review. It took until the World Summit for Sustainable Development in 2002 for aid flows to rise again to the level they had reached in 1990. However, the only new finance came through the UN Global Environmental Facility (GEF), which provided US$1 billion to:

> provide new and additional grants and concessional funding to cover the "incremental" or additional costs associated with transforming a project with national benefits into one with global environmental benefits.
>
> (GEF, 1991)

The 1990s also saw the advancement of globalization, and with it a wave of privatization across much of the world. This was also true in the former Soviet bloc countries as they embraced capitalism and privatized much that had been under state control. This caused significant corruption in these resource-rich sub-national states (Berkowitz, 2001). Developed countries' companies benefitted from globalization as they were able to source cheap labour and raw materials without worrying about environmental or labour laws to the level they had to in their host countries.

There had been an attempt to have a forty-first chapter of Agenda 21 on 'Transnational Corporations and Sustainable Development' (UN Commission on Transnational Corporations [TNC]). In the absence of any control exercised over the activities of TNCs in the 1990s, their impact was often negative in many developing countries. Although they brought some jobs and foreign direct investment, most of the profits were taken out of the host country and many local firms suffered greatly because of TNCs undermining the development of local businesses.

This contributed to the growth of anti-globalization movements, with perhaps the most vivid images coming from the World Trade Organization's Ministerial Conference held in Seattle (1999) which at times seemed to be the site of pitched battles (see Figure 1.2). As the *Sunday Independent* put it:

> The way it has used [its] powers is leading to a growing suspicion that its initials should really stand for World Take Over. In a series of rulings it has struck down measures to help the world's poor, protect the environment, and safeguard health in the interests of private – usually American – companies. 'The WTO seems to be on a crusade to increase private profit

Figure 1.2 Anti-globalization demonstrations in Seattle. Source: Wikimedia Commons/ Steve Kaiser, https://commons.wikimedia.org/wiki/File:WTO_protests_in_Seattle_November_30_1999.jpg

> at the expense of all other considerations, including the well-being and quality of life of the mass of the world's people,' says Ronnie Hall, trade campaigner at Friends of the Earth International. 'It seems to have a relentless drive to extend its power.'
>
> (*Sunday Independent*, 1999)

In the area of debt relief for developing countries, a campaign by NGOs was successful in drawing attention to the fact that much of their national debt had been built up by repressive regimes. As more countries were becoming democratic, views on debt relief and approaching repressive regimes were changing:

> At the same time growing debt in developing countries caused a campaign called Jubilee 2000 which aimed to have the debt for the poorest countries cancelled. It ultimately saw US$130 billion of debt written off for the 35 poorest, mostly heavily indebted countries.
>
> (*Telegraph*, 2014)

During the 1990s governments crafted action plans for areas critical to ensuring movement towards a more sustainable and equitable world. These were the:

- 1990 World Summit for Children *(set twenty-seven goals on health and education for children)*
- 1992 Earth Summit *(see above)*
- 1994 International Conference on Population and Development *(set four goals on education and health)*
- 1995 World Summit on Social Development *(agreed ten commitments in the areas of poverty, employment and fostering social integration)*
- 1995 Fourth World Conference on Women: Action for Equality, Development and Peace *(agreed objectives on poverty, education and training, violence, health, armed conflict, economics, environment, girl-children and media)*
- 1996 Second United Nations Conference on Human Settlements Habitat II *(agreed goals for ensuring adequate shelter for all and enabling more sustainable human settlements)*
- 1996 World Food Summit *(pledged to achieve food security, eradicate hunger and reduce the number of undernourished people)*

At the five-year review of Agenda 21 in 1997 a number of additional chapters were added to it, most significantly the chapter dealing with energy. Although not extremely progressive, it was achieved with the help of governments that had previously been against the inclusion of an energy section.

> The increase in the level of energy services would have a beneficial impact on poverty eradication by increasing employment opportunities and

> improving transportation, health and education. Many developing countries, in particular the least developed, face the urgent need to provide adequate modern energy services, especially to billions of people in rural areas. This requires significant financial, human and technical resources and a broad-based mix of energy sources.
>
> (UN, 1997)

As the millennium was ending there was a recognition that the hopes raised by these UN conferences, summits and action plans had not been realized, and the funds to achieve them had failed to materialize. There needed to be a clearer and more focused agenda if there was to be progress.

Water, energy, food and the Millennium Development Goals

By April 2000 the UN Secretary-General Kofi Annan was worried that preparations for the planned Millennium Summit in September were not going well, so he started to work with the Organization for Economic Cooperation and Development OECD, the World Bank and prominent donor countries on what would become the Millennium Development Goals. Most of the targets were based on ones already agreed by the Development Assistance Committee (DAC) of the OECD. Some had also been suggested by UN conferences or summits in the 1990s. An example of this comes from the Rome Food Summit:

> We pledge our political will and our common and national commitment to achieving food security for all and to an ongoing effort to eradicate hunger in all countries, with an immediate view to reducing the number of undernourished people to half their present level no later than 2015.
>
> (FAO, 1996)

Suggested in 2001, the Millennium Development Goals (MDGs) were not a linear extension of the Millennium Declaration adopted at the 2000 summit. Some principles were lost in translation, including references to human rights. The goals were also not aligned with sustainable development; they represented support for human development. The goals, targets and indicators would be reported on over the next 15 years. The eight initial goals were:

1. To eradicate extreme poverty and hunger
2. To achieve universal primary education
3. To promote gender equality and empower women
4. To reduce child mortality
5. To improve maternal health
6. To combat HIV/AIDS, malaria and other diseases
7. To ensure environmental sustainability
8. To develop a global partnership for development.

On the energy front, in 2000 the United Nations Commission on Sustainable Development (UN CSD) set up a working group to try to develop a new path on renewable energy in preparation for the World Summit on Sustainable Development (WSSD) in 2002. As the UN CSD was unable to develop the commitments further, it was left to the WSSD to see if an additional target on renewable energy could be agreed upon. However, the WSSD, held only a year after 9/11, destroyed any hopes for a new deal between developed and developing countries as governments focused on the new terrorist threats.

Sustainable development stalled with no agreed target on energy. The WSSD host country, South Africa, and the European Union tried to push for a target of at least 5 percent of the world's total energy mix coming from renewables – this failed because of a coalition similar to the one in 1992, between the US and oil states.

The 2002 WSSD's Plan of Implementation underlined the crucial role that agriculture plays in addressing both the needs of a growing population and poverty eradication. It went on to say:

> Sustainable agriculture and rural development are essential to the implementation of an integrated approach to increasing food production and enhancing food security and food safety in an environmentally sustainable way.
>
> (UN, 2002b)

The WSSD also recognized the decline in public sector funding for sustainable agriculture and the need to adopt policies and implement laws that would guarantee "well defined and enforceable land and water use rights, and promote legal security of tenure" (UN, 2002b).

When Germany hosted the Bonn Freshwater Conference a year earlier in 2001, water issues rose in the minds of policy makers. The conference came up with the so-called Bonn Keys.

Box 1.1 Bonn Keys

1 The first key is to meet the water security needs of the poor – for livelihoods, health and welfare, production and food security, and to reduce vulnerability to disasters. Pro-poor water policies focus on listening to the poor about their priority water security needs. It is time now to build on the national and international commitment on drinking water with the determination also to *halve the number of those who do not have access to sanitation.*

2 Decentralization is key. The local level is where national policy meets community needs. Local authorities – if delegated the power and the means, and if supported to build their capacities – can provide for increased responsiveness and transparency in water management, and increase the participation of women and men, farmer and fisher, young and old, town and country dweller.
3 The key to better water outreach is new partnerships. From creating water wisdom, to cleaning up our watersheds, to reaching into communities – we need new coalitions. Energized, organized communities will find innovative solutions. An informed citizenry is the frontline against corruption. New technologies can help; so can traditional techniques and indigenous knowledge. This Bonn stakeholder dialogue is part of the process.
4 The key to long-term harmony with nature and neighbour is cooperative arrangements at the water basin level, including across waters that touch many shores. We need integrated water resource management to bring all water users to the information sharing and decision making tables. Although we have great difficulty with the legal framework and the form agreements might take, there is substantial accord that we must increase cooperation within river basins, and make existing agreements more vital and valid.

The essential key is stronger, better performing governance arrangements. National water management strategies are needed now to address the fundamental responsibilities of Governments: laws, rules and standard setting; the movement from service delivery to the creation and management of an effective legal and regulatory framework. Effective regulatory arrangements that are transparent and can be monitored are the way to effective, responsive, financially sustainable services. Within these we will welcome both improved public sector and private sector delivery arrangements (Germany, 2001).

With ten years between the Dublin Principles and the Bonn Keys, a change in the perception of water was readily apparent. No longer was the issue of water seen merely as an economic good; rather, the focus moves towards integrated water resource management. This was an early approach to the Nexus by the water community. It emphasized the role of governments in being responsible for laws, rules and standard setting, and called for effective regulation to underpin the role of both public and private sector delivery mechanisms, thus learning from the failures of a near-free market approach after Rio 1992. Both Bonn and

Dublin recognized the role of decentralization in delivery to the community, but Bonn underlined the centrality of the roles of women AND local stakeholders to the management of water. Bonn also introduced the concept of water security, which went beyond Dublin's recognition of water as a finite resource. Finally, it also brought the issue of sanitation into the global discourse.

The meeting in Bonn identified the need to include a target on sanitation in the MDGs, and this was picked up by the WSSD. In fact, some people would say that this was the major political outcome, since other proposed targets did not make it to the final text. Another important outcome from WSSD was the call for the UN Secretary-General to begin:

> utilizing the United Nations System Chief Executives Board for Coordination, including through informal collaborative efforts, to further promote system-wide inter-agency cooperation and coordination on sustainable development …
>
> (UN, 2002b)

This resulted in the setting up of three inter-agency bodies: UN Water, UN Energy, and UN Oceans.

The United Nations High Level Committee on Programmes formalized UN Water in 2003. Interestingly, it initially emerged as a top-down response to the WSSD's call, as a decision by the programme managers of the twenty-three concurring UN agencies with a will to coordinate in a development plan paralleling the Bonn Keys but with an emphasis on cooperation and governance. UN Water provides:

> the platform to address the cross-cutting nature of water and maximize system-wide coordinated action and coherence. UN Water promotes coherence in, and coordination of, UN system actions aimed at the implementation of the agenda defined by the Millennium Declaration and the World Summit on Sustainable Development as it relates to its scope of work.
>
> The main purpose of UN Water is, thus, to complement and add value to existing programmes and projects by facilitating synergies and joint efforts, so as to maximize system-wide coordinated action and coherence as well as effectiveness of the support provided to Member States in their efforts towards achieving the time-bound goals, targets and actions related to its scope of work as agreed by the international community, particularly those contained in the Millennium Development Goals and the Johannesburg Plan of Implementation (World Summit on Sustainable Development).
>
> (UN Water, 2014)

Similarly, UN Energy was established as a subsidiary body of the UN Chief Executive Board (UN CEB) in 2004 to help ensure coherence in the United Nations multi-disciplinary response to the World Summit on Sustainable Development (WSSD), and to promote the effective engagement of non-UN stakeholders in implementing energy-related WSSD decisions. Membership consists of senior officials and experts on energy from the commissions, organizations, funds and programmes.

Unlike water and energy, which call on multiple funds, programmes and specialized agencies, the issue of food has a lead UN specialized agency, the UN Food and Agriculture Organization (UN FAO). They work with the other food-related Rome-based agencies: the World Food Programme (WFP) and the International Fund for Agricultural Development (IFAD).

The strange death and rebirth of sustainable development

Many left the WSSD depressed at the state of sustainable development. A summit that had promised high hopes for a new financial deal between developed and developing countries, and which most had expected to adopt targets on the main areas that would help us move in a more sustainable direction, had failed nearly completely.

The reaction to this failure was mixed. Some focused on a reform of the Commission on Sustainable Development into a new two-year cycle (it had previously met annually). A new approach was developed to try to root the Commission in lessons learnt. The first year of the two-year cycle would focus on a review of progress, and the second would be a year of policy changes. Policies would be developed to address challenges and gaps identified. Other people went on to focus on creating 'coalitions of the willing', known as 'Type II outcomes' in the UN bureaucracy, particularly in the area of energy. At Johannesburg the Renewable Energy and Energy Efficiency Partnership (REEEP) was created as such a 'Type II' outcome.

> This Summit will be remembered not for the treaties, commitments, or eloquent declarations it produced, but for the first stirrings of a new way of governing the global commons, the beginnings of a shift from the stiff formal waltz of traditional diplomacy to the jazzier dance of improvisational solution oriented partnerships that may include non-government organizations, willing governments and other stakeholders (type II partnerships).
>
> (World Resources Institute, 2002)

Expressing their frustration at the lack of a target on energy from the WSSD, a number of countries issued the Declaration on Renewable Energy:

> We have adopted, or will adopt, such targets for the increase of renewable energy and we encourage others to do likewise. We are convinced that this will help to implement the necessary policies to deliver a substantial increase in the global share of renewable energy sources. Such targets are important tools to guide investment and develop the market for renewable energy technologies.
>
> (Declaration on Renewable Energy, 2002)

In 2004 the Bonn Renewable Energy Conference again demonstrated the leadership of Germany. One hundred and fifty-four countries attended, producing three outcomes:

> 1 A Political Declaration containing shared political goals for an increased role of renewable energies and reflecting a joint vision for a sustainable energy future that provides better and more equitable access to energy, as well as increased energy efficiency.
> 2 An International Action Program of voluntary commitments to goals, targets and actions within their own spheres of responsibility from governments, international organizations and other stakeholders.
> 3 Policy Recommendations for Renewable Energies that can be of benefit to governments, international organizations and stakeholders as they develop new approaches and political strategies and address the roles and responsibilities of key actors.
>
> (Germany, 2004)

It also called for the establishment of the International Renewable Energy Agency (IRENA).

The Commission on Sustainable Development looked at water and sanitation in its 2004–2005 cycle and realized that little had happened in the two years since WSSD, and therefore agreed to relook at the issue in 2008–2009. As early as 2006 President Mbeki of South Africa was already calling the Johannesburg Plan of Implementation (JPoI) a "forgotten piece of paper." In his speech to the United Nations General Assembly as Chair of G77 he said:

> Precisely because of the absence of a global partnership for development, the Doha Development Round has almost collapsed. Indeed, because the rich invoked, without shouting it, the slogan of an over-confident European political party of the 1960s, and directed this uncaring declaration to the poor of today – "I'm alright, Jack!" – we have not implemented the Monterrey Consensus on Financing for Development, thus making it difficult for the majority of the developing countries, especially those in Africa, to achieve the Millennium Development Goals, and have reduced the Johannesburg Plan of Implementation to an insignificant and perhaps forgotten piece of paper.
>
> (Mbeki, 2006)

The lack of financing was undermining the implementation of actions towards more sustainable development. The reality was that aid flows were going up and would double to US$134.8 billion, or 0.3 percent of gross national income, in the decade following WSSD (OECD, 2014).

Prior to the financial crisis of 2008 it looked like most MDG targets would be achieved, except perhaps some of the global development partnerships (MDG8) and the sanitation target (part of MDG7). However, the 2007 UN

CSD meeting seemed to underline the death of sustainable development when, for the first time, it failed to agree on any policy measures at all. It particularly failed in the sectors of energy and climate change.

In August 2007 the Stockholm International Water Institute (SIWI) and Stakeholder Forum established what would be known as the Water and Climate Coalition. This coalition sought to have water included within the UNFCCC negotiations. At that point no UNFCCC text mentioned water, although it is obvious that energy production uses water (fossil fuel production uses considerable amounts compared with renewable energy production – see Figure 1.3) and similarly water management demands energy. Water is also impacted by climate change, and one particularly important area in this regard is water for food production. Agriculture demands around 70 percent of the water that falls for irrigation (IFAD, 2014), and demand is likely to increase in response to the food demands of an increasing, and increasingly wealthy, population.

Responding to this state of affairs in September 2007, the Brazilian President Luiz Inácio Lula da Silva addressed the UN General Assembly at the opening of its annual leaders' week, making the case for a new conference on sustainable development. He indicated the necessity for a summit that would "*address the persistent problem of global inequality*", asserting that "*social equality is our best weapon against the planet's degradation.*" He stressed the need to reorder international priorities around an agenda that supported and favoured "social justice." Recalling Rio's efforts to preserve and protect "*our common heritage*", President Lula stated that this heritage could only be "*salvaged through a new and more balanced distribution of wealth*" (Lula, 2007).

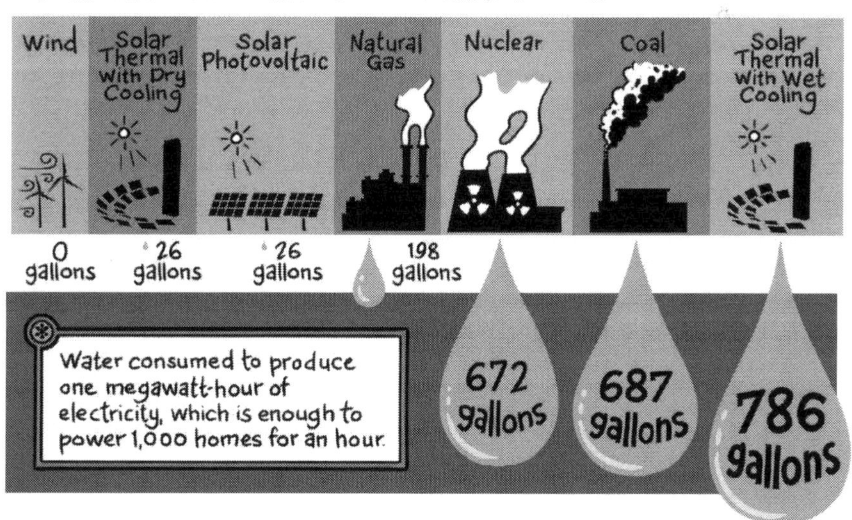

Figure 1.3 Water use by power plants. Source: National Renewable Energy Laboratory. Illustration by Andy Warner.

This paved the way for Rio+20, the 2012 United Nations Conference on Sustainable Development, and the rebirth of sustainable development which has now become the main organizing principle for the development of universal goals and targets. This rebirth brought together the human development (MDG) and sustainable development movements into a common UN framework.

In June 2008 the IPCC published Climate Change and Water: IPCC Technical Paper VI. This says:

> Freshwater-related issues are critical in determining key regional and sectoral vulnerabilities. Therefore, the relationship between climate change and freshwater resources is of primary concern to human society and also has implications for all living species.
>
> (IPCC, 2008)

At the UNFCCC negotiations in Cancún in December 2010, the Water and Climate Coalition and six governments made a call, which was put forward by both Ecuador and Sudan and was further supported by Syria, Chile, El Salvador and Sierra Leone. The call proposed the inclusion of water on the agenda for the next meeting of the body which provides scientific and technical advice to the climate convention – the SBSTA. The motion was put forward by the representative of Ecuador, Undersecretary Tarsicio Granizo:

> Ecuador feels that water should be addressed more prominently at the climate change negotiations. Climate change impacts will primarily be felt through water, and the way we manage our water will be critical to our resilience. This issue has long been neglected at the intergovernmental level and we would welcome the opportunity to discuss it at greater length in the future.

The representative of Ecuador also noted that his country has recognized that water is a human right. Ambassador Lumumba Di-Aping, who represented Sudan, said:

> Sudan understands the challenges facing countries through the impacts of climate change on water resources. Yet so far we have not addressed this issue adequately through the climate change negotiations. Getting water on the agenda will help us to identify these challenges and propose solutions.

The introduction of the Nexus

With the financial crisis of 2008–2011, a clear link between food and energy was made. As energy prices went up, so did food prices, since food needs to be grown, manufactured and transported with an associated high use of energy. The average price of energy during 2000–2012 was 183.6 percent higher than the average price during 1990–1999; in the same period the World Bank global food price index increased by 104.5 percent, at an average annual rate of 6.5 percent.

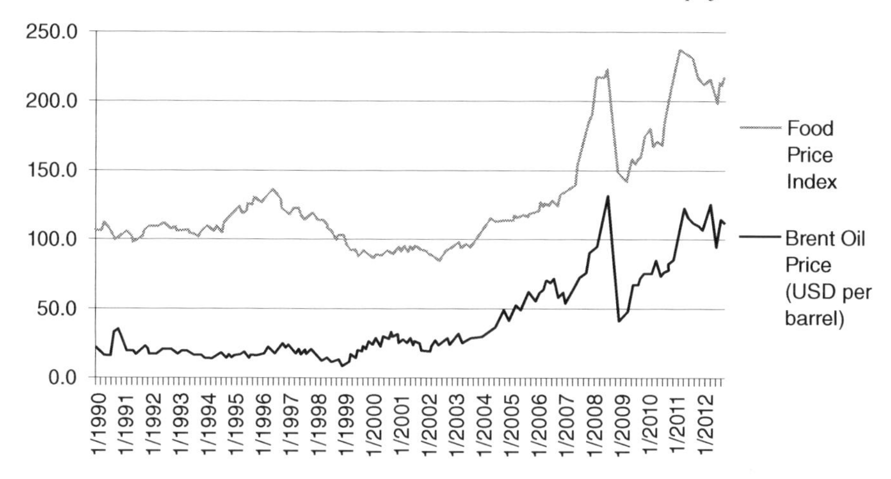

Figure 1.4 FAO Food Price Index, EIA Brent Oil Price.

In 2009 the German government hosted an event at the World Water Week in Stockholm to discuss what their contribution to the proposed Sustainable Development Conference in 2012 should be. At the workshop there was support for addressing the interlinkages between water, food and energy use, as well as the impact of increased urbanization. This was a tall order considering that links between sectors had been such a difficult area for the CSD to address. The German government returned to the following year's Water Week with a proposal for a 'Nexus Conference'. With additional feedback secured, they set up an International Advisory Board which would meet from late 2010 until the conference in November 2011.

Underpinning the conference was a background report produced by the Stockholm Environment Institute. It suggested a number of drivers that were impacting the Nexus. These included the expected population growth with an additional billion people by 2030, and increased urbanization with 54 percent of the world living in urban areas by 2030 and 70 percent by 2050; compared with 34 percent in 1960. While the changing nature of economic growth, particularly in China and India, would further these factors, the report estimated that the world would require an additional 40 percent of energy compared to 2011, create a 30–50 percent increase in food demand, and would contribute to a 40 percent gap (between availability and demand) in water resources (see Figure 1.5).

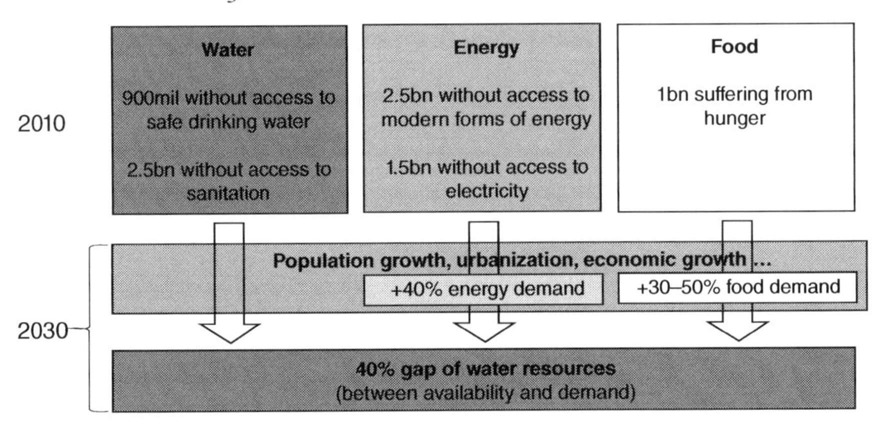

Figure 1.5 At the tipping point.

The Rio+20 outcomes

As Heads of State started to arrive in Rio in June 2012 it looked like the work that had started five years earlier would not deliver a new path to sustainable development. The last preparatory meeting for Rio+20, held in New York from May 29 to June 2 2012, had ended with a negotiation text that was far from ready for Heads of State to assess. The eighty-page revised draft produced by the co-chairs had been reworked by two working groups and over twenty issue-specific contact or 'splinter' groups. As country representatives landed for the remaining negotiations and the conference itself there were only 70 paragraphs agreed, with 259 containing bracketed text. There were areas of divergence on important issues such as finance and the green economy, the reform of the CSD and the UNEP, technology transfer, climate change, oceans, sexual and reproductive health rights and food and agriculture; as well as the process for the establishment of Sustainable Development Goals and the means of implementation.

Again, the Brazilian government adopted a critical leadership role, taking on the negotiations for the final preparatory session and concluding the negotiations in a record three days. The outcome had a number of very clear decisions.

On Institutional Framework for Sustainable Development

- UNEP would become a 'universal body' in the UN, falling short of becoming a UN Agency but possibly preparing the ground for that to happen in the future.
- The UN CSD would close down and be replaced by a new 'body', the High Level Political Forum, to sit between the United Nations Economic and Social Counsel and the UN General Assembly with

universal membership and meeting at the level of Heads of State and Government every four years.

In the area of Sustainable Development Goals, finance and technology transfer

- An open-ended working group of thirty members to develop the Sustainable Development Goals was agreed upon.
- An Intergovernmental Expert Committee on Sustainable Development Finance was agreed upon.
- Relevant UN agencies were tasked with identifying options for a technology facilitation mechanism to be presented to the 67th UN General Assembly.

A backdrop to the conference was the growing realization that the Water, Energy and Food Security Nexus would be one of the major challenges in achieving universal access to these basic services without compromising planetary boundaries in the coming years. This Nexus was reflected in the sector commitment areas. In particular, the Food Security and Nutrition and Sustainable Agriculture section of 'The Future We Want' (the output of Rio+20) aimed to:

> f) Reaffirm the necessity to promote, enhance and support more sustainable agriculture, including crops, livestock, forestry, fisheries and aquaculture, that improves food security, eradicates hunger, and is economically viable, while conserving land, water, plant and animal genetic resources, biodiversity and ecosystems, and enhancing resilience to climate change and natural disasters (TFWW, V.111).
>
> (UN, 2012)

While the Nexus was not reflected in the Energy section of 'The Future We Want', it was mentioned in the section on climate change:

> a) Reaffirming that climate change is one of the greatest challenges of our time, and we express profound alarm that emissions of greenhouse gases continue to rise globally. We are deeply concerned that all countries, particularly developing countries, are vulnerable to the adverse impacts of climate change, and are already experiencing increased impacts including persistent drought and extreme weather events, sea level rise, coastal erosion and ocean acidification, further threatening food security and efforts to eradicate poverty and achieve sustainable development. In this regard, we emphasize that adaptation to climate change represents an immediate and urgent global priority (TFWW, V.190).
>
> (UN, 2012)

However, perhaps the place where the Nexus concept was most reflected was in the Water and Sanitation section:

> a) Promote the reuse of treated wastewater and water harvesting and storage (TFWW, V.109).
> b) Recognize that water is at the core of sustainable development as it is closely linked to a number of key global challenges (TFWW, V.119).
> c) Commit to the progressive realization of access to safe and affordable drinking water and basic sanitation for all, as necessary for poverty eradication, women's empowerment and to protect human health, and to significantly improve the implementation of integrated water resource management at all levels as appropriate (TFWW, V.120).
> d) Reaffirm commitments regarding the human right to safe drinking water and sanitation, to be progressively realized for our populations with full respect for national sovereignty (TFWW, V.121).
> e) Recognize the key role that ecosystems play in maintaining water quantity and quality and support actions within respective national boundaries to protect and sustainably manage these ecosystems (TFWW, V.122).
> f) Adopt measures to address floods, droughts and water scarcity, addressing the balance between water supply and demand, including, where appropriate, non-conventional water resources, and to mobilize financial resources and investment in infrastructure for water and sanitation services, in accordance with national priorities (TFWW, V.123).
> g) Stress the need to adopt measures to significantly reduce water pollution and increase water quality; as well as significantly improve wastewater treatment and water efficiency and reduce water losses. In order to achieve this, stress the need for international assistance and cooperation (TFWW, V.124).
>
> (UN, 2012)

While the media was very critical of Rio+20, it did in fact create some strong foundations to rebuild sustainable development. It strengthened, at the global level, the organizations required to monitor and hold accountable those responsible for future delivery. This applied not only to Rio+20 but also to whatever new development goals would be agreed on in 2015. It also set up the mechanism to enable governments to agree what those goals should be over a span of three years, and how they might be financed.

By the end of Rio a set of useful definitions was established. Water security had been defined in the Millennium Development Goals (MDGs) as "access to safe drinking water and sanitation", and both of these had been recognized as human rights in the interim. While they are not part of most water security definitions as yet, availability of, and access to, water for other human and ecosystem uses is also very important from a Nexus perspective.

Energy security had been defined as "access to clean, reliable and affordable energy services for cooking and heating, lighting, communications and productive uses" (United Nations, UN Secretary-General's Advisory Group on Energy and Climate Change), and as "uninterrupted physical availability [of energy] at a price which is affordable, while respecting environment concerns."

Food security is defined by the Food and Agricultural Organization (FAO) as "availability and access to sufficient, safe and nutritious food to meet the dietary needs and food preferences for an active and healthy life" (Rome Declaration, 1996). Adequate food has also been recognized as a human right.

The development of the Sustainable Development Goals

From Solo in June 2011, when Colombia and Guatemala first put forward the concept of SDGs, to September 2015 it required over four years of work to change course from a development agenda focused solely on developing countries to a universal sustainable development agenda relevant to, and demanding action of, ALL countries. That path was not easy. Initial opposition by governments relevant to, and demanding action of, development agencies and northern development NGOs had to be overcome.

Colombia had advocated for the need for a comprehensive, sector-connecting approach to the targets. They expressed concern that a sectoral approach would not enable the "transformational agenda" needed, and nor would it address the reality that areas like health, water, energy and food impact substantially on each other and so need to be addressed together. Early on they gave an example of what they thought we should be looking for in the area of a health goal based on the MDG targets. It compared: Target 1C: Hunger; Target 4A: Under-5 mortality rate; Target 5A: Maternal mortality ratio; Target 5B: Reproductive health; Target 6A: HIV/AIDS; Target 6B: HIV/AIDS treatment; Target 6C: Malaria and other; Target 7C: Sanitation and drinking water; and Target 7D: slum dwellers.

Parallel to the Rio+20 processes the UN Secretary-General had been thinking of what he could lead on in terms of new MDGs, and set in motion a number of initiatives:

- A High Level Panel of Eminent Persons on the Post-2015 Development Agenda – co-chaired by President Susilo Bambang Yudhoyono of Indonesia, President Ellen Johnson Sirleaf of Liberia and Prime Minister David Cameron of the United Kingdom, and including leaders from civil society, the private sector and sub-national governments.
- The Sustainable Development Solutions Network under the leadership of Jeffrey Sachs.
- Over 120 National Consultations mostly coordinated by UNDP.
- Global thematic consultations compiling stakeholders' inputs around current and emerging challenges in respect to eleven defined

substantive issues. These consultations were intended to spark a broad coalition for change around specific themes, benefiting from the lessons learned in the implementation of the MDGs and identifying new opportunities for eradicating poverty and enhancing human development. The eleven consultations were on Inequalities, Governance, Growth and Employment, Health, Education, Environmental Sustainability, Food Security and Nutrition, Conflict and Fragility, Population Dynamics, Energy, and Water.

- A Sustainable Development Goal Open Working Group to prepare for a future set of goals.

The SDG Open Working Group turned out to be very popular, and against the thirty proposed seats, seventy countries decided they wanted to take part. A deal was reached to establish a 'buddy system'.

Box 1.2 Members of the SDG Open Working Group

African Group

Algeria/Egypt/Morocco/Tunisia
Ghana
Benin
Kenya
United Republic of Tanzania
Congo
Zambia/Zimbabwe

Asia-Pacific Group

Nauru/Palau/Papua New Guinea
Bhutan/Thailand/Viet Nam
India/Pakistan/Sri Lanka
China/Indonesia/Kazakhstan
Cyprus/Singapore/United Arab Emirates
Bangladesh/Republic of Korea/Saudi Arabia
Iran (Islamic Republic of)/Japan/Nepal

Latin American and Caribbean Group (GRULAC)

Colombia/Guatemala
Bahamas/Barbados
Guyana/Haiti/Trinidad and Tobago
Mexico/Peru

Brazil/Nicaragua
Argentina/Bolivia (Plurinational State of)/Ecuador

Western European and Others Group (WEOG)

Australia/Netherlands/United Kingdom of Great Britain and Northern Ireland
Canada/Israel/United States of America
Denmark/Ireland/Norway
France/Germany/Switzerland
Italy/Spain/Turkey

Eastern European Group

Hungary
Belarus/Serbia
Bulgaria/Croatia
Montenegro/Slovenia
Poland/Romania

There were some interesting buddy groups with India/Pakistan/Sri Lanka, China/Indonesia/Kazakhstan, Iran/Japan/Nepal, Cyprus/Singapore/United Arab Emirates and Italy/Spain/Turkey. Importantly, this approach broke up traditional negotiating blocks such as the EU and G77.

The SDG Open Working Group met thirteen times over a period of a year and absorbed not only inputs from the national consultations, the thematic consultations and the High Level Panel of Eminent Persons, but also the first ten sessions allowed additional input on the major areas that had been identified in section 5 of the Rio+20 document, 'The Future We Want' and the High Level Panel of Eminent Persons.

The President of the 66th UN General Assembly, Ambassador John Ashe of Antigua and Barbuda, added a number of Interactive Dialogues on areas that were relevant to the development of the SDGs. Such dialogue, on energy and water in February 2014, had a particular focus on Nexus discussions. In parallel, the second Nexus Water–Energy–Food–Climate Change Conference was to be held in Chapel Hill, North Carolina on March 5–8 2014. The Conference draft declaration was presented to governments at the dialogue for discussion and input.

The Nexus from Bonn to Chapel Hill to the Sustainable Development Goals

As mentioned above, the first Nexus Conference was held in Germany under the sponsorship and leadership of the German government. In 2013 they were still carrying the torch for the Nexus but were not intending to organize another conference. The UNC decided to pull together an international advisory board (IAB), which included one of the co-chairs from the first conference, the former Rwanda Energy Minister Albert Butare. The conference had strong UN and intergovernmental support from UNDESA, UNDP, IFAD, UN Habitat, UNESCO, the World Bank and IRENA. The IAB proposed and guided the conference towards realizing three objectives: to re-establish a political space for the Nexus to grow; to provide input to the SDGs from a Nexus perspective; and to start to develop materials that explain the Nexus.

The Water Institute at the University of North Carolina (UNC) organized the 2014 Nexus Conference. The Institute is run by Jamie Bartram, who prior to this appointment was water expert at the World Health Organization (WHO) and chaired the multi-stakeholder platform on Water and Climate. Jamie co-directed the conference with Felix Dodds, Fellow at the Global Research Institute (GRI) and one of the coordinators of the Water and Climate Coalition. Felix had also been a member of the 2011 Nexus International Advisory Board and put together the multi-stakeholder input to the conference.

Over three hundred people attended the conference representing governments, intergovernmental organizations, academics, NGOs, industry and other stakeholders. It successfully developed a final declaration from the co-directors of the conference, Jamie Bartram and Felix Dodds, which was shared with the governments in New York who were negotiating the SDGs.

Box 1.3 Building integrated approaches into the Sustainable Development Goals

A Declaration in the name of the co-directors of the Nexus 2014: Water, Food, Climate and Energy Conference held at the University of North Carolina at Chapel Hill, March 5 to 8 2014

As the Open Working Group formulates its recommendations for future Sustainable Development Goals (SDGs), we urge its members and chairs to consider that:

1 The SDGs are an opportunity to articulate the promise of a well-executed sustainable development agenda in a set of goals, targets and indicators that are specific, universal and flexible and capable of being achieved by 2030.

2 The SDGs must catalyze sustainable development in all countries, parts of society and sectors of the economy by reanimating institutional structures and initiatives at every governance level to advance human well-being and protect the integrity and resilience of the planet's natural systems.
3 The SDGs must express a balanced determination to achieve both poverty eradication and other development objectives, and to move decisively towards more equitable and sustainable patterns of consumption and production throughout the world.
4 The SDGs must push national and supranational entities to re-design institutional arrangements to encourage partnerships that span sectors and incorporate business, citizens and other stakeholders.
5 The SDGs must highlight the added value of combining the sustainability and poverty reduction agendas rather than casting sustainability as a necessary constraint. Potential examples of this approach include the concept of irreversible poverty eradication and of maximizing the sustainable yield from the oceans for human nutrition.
6 Governments must engage all stakeholders in the sustainable development agenda-setting process. The business community at all levels should be an active partner in order to further promote improved integrated resource management, more effective multi-sector collaborations, more equitable pricing structures, and more inclusive business models. An effective sustainable development agenda will encourage companies to overcome resource mismanagement and go beyond it to the opportunities created by proper resource management.
7 Governments should establish national multi-stakeholder bodies chaired by Heads of State and reporting to each country's respective legislative body as well as the High Level Political Forum at the United Nations in order to help secure implementation of the SDGs and ensure an integrative approach.
8 Governments have a role to play in correcting market failures. They need to ensure the integrity of financial markets and realign incentives and behaviours for a healthier financial system that accounts for sustainability issues. They must also support regulatory evolution to support new mechanisms and create market environments that will shift capital towards sustainable development and supporting the eradication of poverty.

9 The scientific community must undertake and deliver interdisciplinary and transdisciplinary research that identifies new policy approaches and grounds them in robust science, defining what is and is not a cross-cutting issue, and providing policy makers with evidence and tools that match their needs.
10 Social entrepreneurs need a funding platform for innovative projects that advance integrated action on cross-sector issues. Such a platform should be paired with experienced practitioners to serve as mentors.
11 Sustainable cities depend on sustainable countrysides and rural transformation. Local and regional governments must respond not just to issue linkages but also to spatial linkages between societies, ecosystems and economies. Urban and rural landscapes and human settlements are inextricably linked and policy-making must reflect this.
12 Ensuring global food security while also eradicating rural poverty requires investing in the empowerment of the five hundred million smallholder farms worldwide to sustain and enhance the provision of goods and services from production systems through agro-ecological and good animal welfare practices, as well as through ensuring and enabling environments for their security of tenure and investment, with particular attention to achieving sustainable management of soils, water, energy and biodiversity. Governments should also eliminate harmful agriculture subsidies.
13 The development community should work with the agriculture community to empower farmers of all scales to sustainably produce healthy, diverse and nutritious food for consumers in ways that reinforce rural and urban livelihoods, mitigate the adverse impacts of contemporary agricultural practices, and facilitate adaptation and resilience in the face of changing climatic conditions.
14 Ensuring global food securities requires a more sustainable and economically viable agriculture based on a proper and efficient balance between plant- and animal-based productions. Such a balance would also enhance the nutritional value of global food consumption, restore soil fertility, protect water sources and plant and animal genetic diversity, promote biodiversity, and enhance ecosystem services and resilience to climate change and natural disasters.

All levels means all levels of government, local, subnational and national

In July 2014 the proposed SDGs were agreed among the governments at the Open Working Group, and though there was some evidence of Nexus targets in the area of food and water there were none in energy:

> 2.4 (Goal 2: End Hunger, achieve food security and improved nutrition, and promote sustainable agriculture): by 2030 ensure sustainable food production systems and implement resilient agricultural practices that increase productivity and production, that help maintain ecosystems, that strengthen capacity for adaptation to climate change, extreme weather, drought, flooding and other disasters, and that progressively improve land and soil quality.
>
> (UN, 2014)

> 6.4 (Goal 6: Ensure availability and sustainable management of water and sanitation for all): by 2030, substantially increase water-use efficiency across all sectors and ensure sustainable withdrawals and supply of freshwater to address water scarcity, and substantially reduce the number of people suffering from water scarcity.
>
> (UN, 2014)

Integrated targets had suffered during negotiations as governments attempted to reduce the overall number of targets. In the end the agreement contained 17 SDGs and 169 targets. The outcome document was discussed at the 67th UNGA on September 11 with the decision:

> ... the proposal of the Open Working Group on Sustainable Development Goals contained in the report shall be the main basis for integrating Sustainable Development Goals into the post-2015 development agenda, while recognizing that other inputs will also be considered, in the intergovernmental negotiation process at the sixty-ninth session of the General Assembly.
>
> (UN, 2014)

Box 1.4 Sustainable Development Goals Open Working Group report

1. End poverty in all its forms everywhere
2. End hunger, achieve food security and adequate nutrition for all, and promote sustainable agriculture
3. Attain healthy lives for all at all ages
4. Provide equitable and inclusive quality education and life-long learning opportunities for all
5. Attain gender equality, empower women and girls everywhere

6 Secure water and sanitation for all for a sustainable world
7 Ensure access to affordable, sustainable and reliable modern energy services for all
8 Promote strong, inclusive and sustainable economic growth and decent work for all
9 Promote sustainable industrialization
10 Reduce inequalities within and among countries
11 Build inclusive, safe and sustainable cities and human settlements
12 Promote sustainable consumption and production patterns
13 Promote actions at all levels to address climate change
14 Attain conservation and sustainable use of marine resources, oceans and seas
15 Protect and restore terrestrial ecosystems and halt all biodiversity loss
16 Achieve peaceful and inclusive societies, rule of law, effective and capable institutions
17 Strengthen and enhance the means of implementation and global partnership for sustainable development.

Present Nexus initiatives

The Nexus approach is not as integrated into the proposed SDGs as might have been hoped. As negotiations towards the Post-2015 Development agenda continue up to its adoption at the UN Special Summit in September 2015, it will be possible to again raise the issue and seek the insertion of integrated targets in the final agreement.

For each target there will be indicators. Indicators also offer an opportunity for the development of a Nexus approach within sector-specific targets. At the time of writing this chapter, it is unclear when the indicators will be agreed upon. The most sensible approach would be to let the UN Statistic Commission agree on the basket of indicators in 2016, and there is much support for this approach. The Statistical Commission membership is governmental, made up of experts from the national statistical departments.

Since the UNC organized the 2014 Conference there has been an increase in Nexus gatherings and discussions. This increased activity has led to the United Nations University organizing a conference and the OECD a two-day workshop. Such conversations have set up further outputs relating to this concept.

Food, Environment, Energy, Water Academic Network

> The global food price shock of 2007–08, for example, raised the number of undernourished and those in extreme poverty by more than 100 million

> people, and much higher wheat prices in 2010–11 might have contributed to the 'Arab Spring'. The recent food crises were, in part, caused by high energy prices and misguided policies to promote biofuel production to improve energy security, but which in fact raised emissions and reduced the land and water available for food production. This shows the importance of linking decisions across food, energy, environment and water systems.
>
> (FE2W, 2014)

In 2014 a new academic network (see Chapter 4) was launched. The Food, Energy, Environment and Water (FE2W) Network works with decision-makers to improve their understanding of systemic risks and how to manage shocks across these systems. Their approach is founded on collaboration and has an emphasis on poverty reduction, sustainable livelihoods and the need to maintain critical ecosystem services. They aim to engage with people who make decisions, from farmers to policy makers to consumers, and to recommend actions that result in improved long-term outcomes.

In the first operational phase, the Network proposes to focus on six, water-defined regions: the Colorado Basin (United States), the Ganges–Brahmaputra–Meghna Basin (South Asia), the Mekong Basin (South-East Asia), the Murray–Darling Basin (Australia), the Nile Basin (East Africa), and the Volta Basin (West Africa). The Network has identified three objective areas:

1 UNDERSTAND RISKS: to the security of food, energy, environment and water resources. The linkages across systems mean that a narrowly focused approach to fixing one problem often exacerbates stresses elsewhere. For example, subsidized electricity may drive excessive groundwater pumping that undermines water, food and energy security. The Network notes that a holistic approach to food, energy, environment and water informs understanding of the potential costs, benefits and feedback effects associated with different management decisions.
2 ENGAGE DECISION-MAKERS: on developing, adapting, and implementing decision frameworks that consider food, energy, environment and water as integrated systems. FE2W proposes to help decision-makers who are confronted by risks and uncertainties to tackle complex problems. Its decision frameworks are intended to provide a pathway to practical, relevant responses that meet specific goals and respond to inter-system risks.
3 ENABLE ACTIONS: by decision-makers, whether farmers or water ministers, that result in improved long-term outcomes for people and the natural resources they depend on. FE2W asserts that its knowledge and experience allows it to support decision-makers in understanding the risks and outcomes across food, energy, environment and water systems, helping them to balance trade-offs, reduce risks, and accommodate uncertainty. It seeks 'win–win' solutions and approaches so that local and national level projects can be scalable across different locations and circumstances and generate improvements to people's lives.

International Water Management Institute (IWMI)

> The availability of water is central to the future security of food, energy, domestic and industrial water supply and the environment. This relationship is the 'nexus' between a set of competing demands and interactions. Producing more food and energy, and having sufficient water for our fast growing population – projected to reach 9.5 billion by 2050 – means managing water, food and energy differently.
>
> (Bird, 2014)

(See Chapter 15) The IWMI has been taking a leading role in promoting an integrated approach since its creation over thirty years ago. The focus and approach to water management has changed over those thirty years as described in this chapter. The IWMI has also broadened its approach to include "water management, sustainable solutions to poverty, food security and environmental degradation" (Bird, 2014).

Its publication *Setting and Achieving Water Related Sustainable Development Goals* emphasises that a holistic approach is needed.

Corporate Engagement, SABMiller and WWF

A number of companies are tackling the Nexus. Some examples of companies that have made explicit references to it are Pepsi, Coke, Unilever, AECOM and SABMiller. Others may be doing this without adopting a Nexus vocabulary.

In 2014 SABMiller and WWF produced a report entitled *The Water–Food–Energy Nexus: Insights into resilient development*. The initial corporate leadership in the Nexus has come from companies that have a link between water and food or drink production. They see it as a critical question for the actual viability of their company. Pepsi became the first company to recognize water as a basic human right:

> In March 2013, we announced a new goal to provide access to safe water to an additional 3 million people over the next 3 years through continued partnerships with water organizations, doubling the company's original goal. As one of the first major consumer product companies to endorse the United Nations Human Right to Water.
>
> (Pepsico, 2014)

Building on this and addressing the more integrated approach required by the Nexus was an obvious next step for SABMiller and WWF (see Chapter 16). Clearly, SABMiller needs to ensure good quality drinking water and sustainable agricultural products, since these are critical for a brewing company. This also requires them to be a 'responsible company' in the societies where they operate. As water becomes more scare, companies that are engaging with local communities in reducing water use will see their ability to operate remain secure. Conversely, companies that have not engaged in this may find their ability to operate challenged by local communities. There are already many examples of water-related conflicts. The map below (Figure 1.6) displays nearly two thousand incidents involving conflict and collaboration over shared river basins.

SABMiller has been working with WWF for a number years on water-related issues, including work on water footprints, and the company is looking to adopt a similar approach to carbon offsets for water.

> Water footprinting is becoming a popular way of understanding the total water input to consumer products such as beverages, food and clothes. Just as the carbon footprint concept has assisted businesses and consumers in understanding the level of greenhouse gas emissions created by their activities, so water footprinting is creating awareness of how and where this precious resource is used.
>
> (SABMiller and WWF International, 2009)

Water-demanding issues will increase as climate change, urbanization, population growth and economic activity all progress in developing countries. Some will extract it, some will pollute it, and some will demand its institutional functions. The challenge is to ensure that water management is done in a way that is as sustainable as possible. The increased risk of resource scarcity can become a shared risk if the approach to core resources changes and becomes more efficient – for example, where trade-offs are needed, they should be undertaken with the involvement of the communities and sectors causing and affected by the uses.

SABMiller and WWF research concluded that:

> the most resilient economic systems combine robust infrastructure, flexible institutions and functioning natural capital. The case studies propose areas in which policymakers have particular levers for responding to nexus challenges in order to bring about resilience. They suggest policymakers should:
>
> - Integrate all aspects of development planning, in particular ensuring that water, energy and agricultural sector planning are not done in isolation, but consider how each can contribute to the resilience of the others;
> - Design institutions for resilience, in ways that strengthen cooperation and coordinated decision-making;
> - Use economic and regulatory instruments to strengthen the incentives and requirements for building resilience into water, food and energy systems;
> - Use trade, regional integration and foreign policy to manage nexus trade-offs more effectively, and contribute further to resilience at both country and global levels.
>
> (SABMiller, 2014)

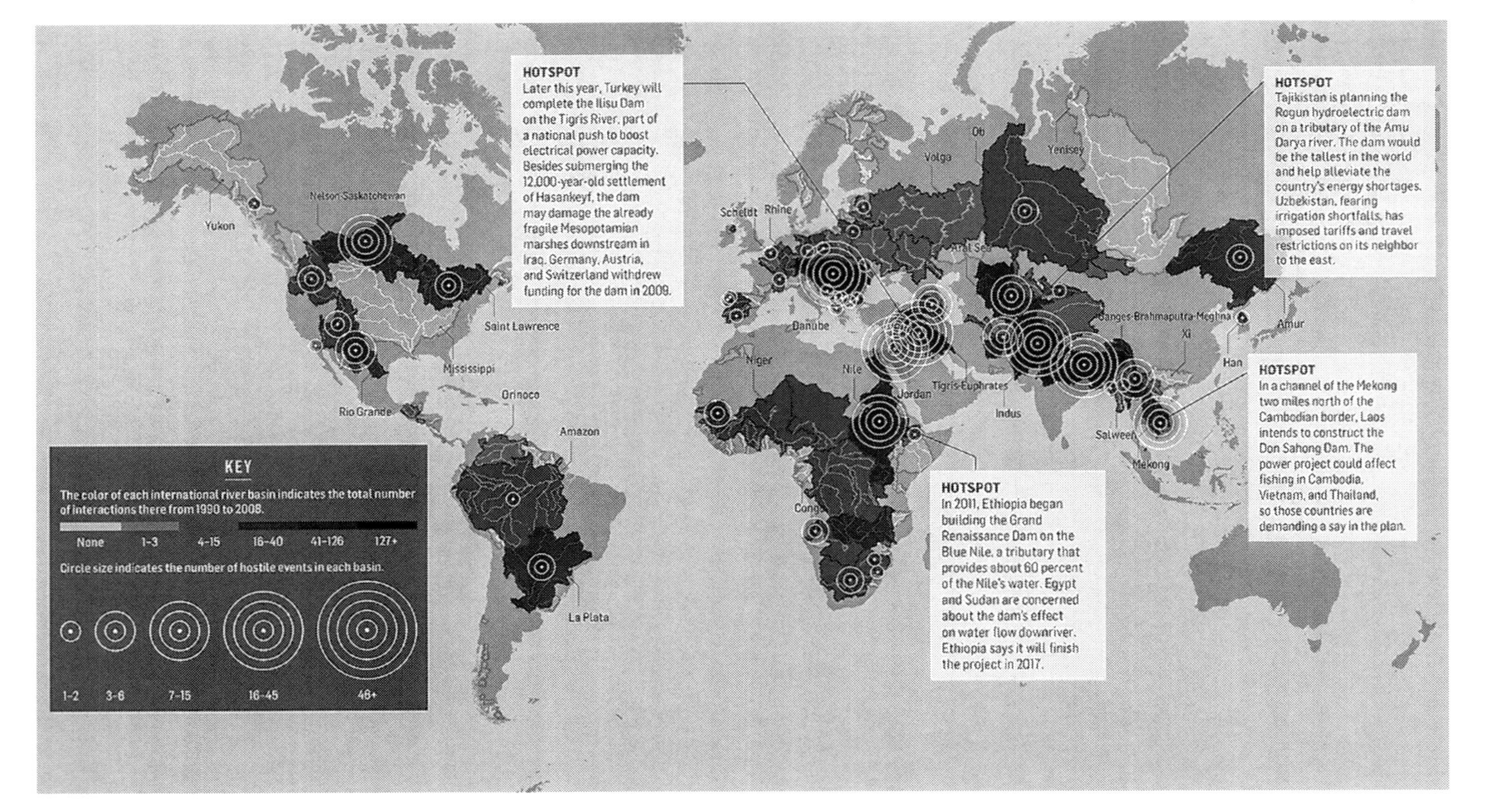

Figure 1.6 Where will the world's water conflicts erupt? Data visualization by Pitch Interactive; river locations courtesy of The Global Runoff Data Centre, 56068 Koblenz, Germany.

ICLEI Local Governments for Sustainability

During 2014, with support from German Development Cooperation (GIZ on behalf of BMZ), ICLEI developed a report entitled '*Operationalizing the Urban NEXUS: Towards resource-efficient and integrated cities and metropolitan regions.*' It looks at how to design sustainable urban development solutions, looking for 'synergies between sectors, jurisdictions and technical domains.'

Much of the real-world decision-making required to tackle the issues highlighted by a Nexus perspective will challenge established sector thinking and will have to look at tradeoffs and how to deal with divided responsibilities between different parts of government, as well as the vertical relations between local, regional and national jurisdictions. The Urban NEXUS Development Cycle, elaborated by ICLEI (Figure 1.7), is a process for translating integrated planning objectives into policies, projects, systems, and places.

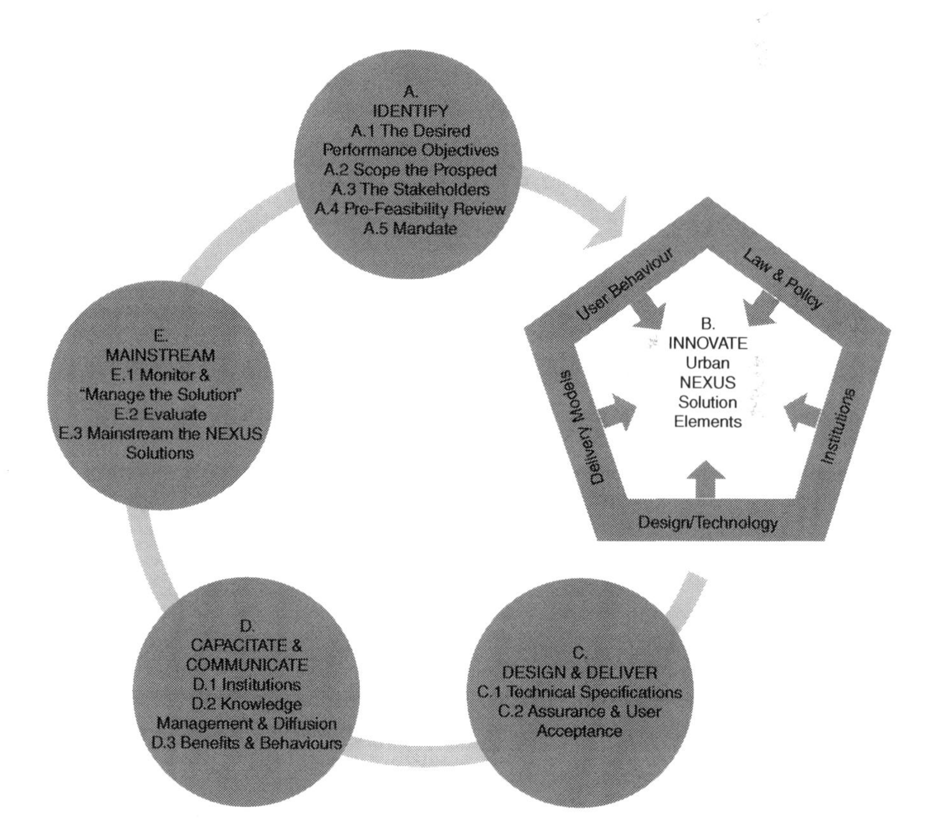

Figure 1.7 The Urban Nexus Development Cycle – a process for translating integrated planning objectives into policies, projects, systems and places. Source: GIZ and ICLEI, 2014.

Box 1.5 The Urban Nexus Development Cycle

1 IDENTIFY. The four primary Urban Nexus Practice Objectives – increased systemic effectiveness, increased demand-driven suitability and customization, increased productive efficiency, and increased resilience and adaptive capacity – are identified and adapted locally. To achieve the strategic objectives, stakeholders then identify the local Urban Nexus integration prospects. Prospects for building synergies can be found in five Areas of Integration:
   - Integration across scales of the built environment, infrastructures, local and regional supply chains and resource cycles, and policies and operations of local, regional, sub-national and national jurisdictions;
   - Integration of systems of resource extraction and power generation, food cultivation, processing, manufacture, resource supply and waste management etc. by establishing cascades and cycles of resources between systems;
   - Integration of services and facilities to avoid the underutilization of valuable fixed assets by integrating services and facilities conventionally separated by sectoral functions;
   - Integration across silos consolidating institutional interests and managerial and professional 'silos' arising from the organization of urban areas and systems into separate jurisdictions, utilities, and departments; and finally
   - Integration of social relations and behaviours to enable the engagement of all stakeholders in the above integration dimensions, and counter legacies of cultural, social, and political division.
2 INNOVATE. The identified stakeholders collaborate in a structured innovation process to develop a set of politically, institutionally and economically viable measures in the Areas of Urban Nexus Innovation, spanning the range of law and policy, design and technology, delivery models and financing, communications and changing user behaviours, and institutional design and development.
3 DESIGN and DELIVER. The design and delivery of the solution includes prototyping and piloting it in a real-world operating environment. This study included the implementation of two such pilot projects in Nashik, India and Dar es Salaam, Tanzania. These cities took first steps in implementing an Urban Nexus approach in an exemplary way. In the limited duration of the pilot projects, the

Urban Nexus brought together a wide range of stakeholders who had never before sat together at one table, thus generating a new 'institutional nexus'. They collaboratively designed and implemented innovative solutions and programmes for optimizing water, energy and land resources in peri-urban agricultural practices (Nashik), and improving the learning environment at two municipal schools while installing integrated energy-efficient technologies, rainwater catchment and vertical food production systems (Dar es Salaam) to demonstrate the benefits of Urban Nexus thinking to local communities and government officials.

4 COMMUNICATE and CAPACITATE. The three main areas of capacity building typically required to establish a new solution are: training operational staff on managing their parts of the solution; encouraging behavioural change and building required skills among beneficiaries; and enabling the relevant institutions to establish a systematic process for introducing and supporting it in new locations or facilities.

5 UPSCALE and MAINSTREAM the Urban Nexus. Mainstreaming in many cases is a matter of designating or creating an entity that specializes in the scaling of all the unique aspects of the given Urban Nexus solution; an entity with the capacity to address location-specific problems and to 'manage the solution' within different contexts. (GIZ and ICLEI, 2014)

Germany and other governments

> With increasing population pressure, it has become apparent that 'silo' approaches have reached their limits in dealing with a complex urban environment, characterized by multiple potential resource conflicts between various urban sectors (water, sanitation, wastewater, solid waste, transport, energy, agriculture, amongst others). The NEXUS approach allows containing these potential conflicts or even creating synergies by, for example, promoting increased productivity of resources and resource use efficiency, in line with circular economy principles.
>
> (Eid, 2014)

Germany has sustained leadership in the area of the Nexus. It has organized Nexus events at conferences and UN-related events, and has funded research and projects that focus on the Nexus. Recently the government of the Netherlands has set up a Nexus Office in their Foreign Ministry, and other governments are looking to follow suit in the coming years.

GRACE Communications Foundation

Some NGOs, like the GRACE Communications Foundation in the United States, have managed to express apparently complex and sometimes opaque Nexus concepts in clear comprehensible terms that appeal to direct action. The front page of GRACE Communications Foundation's website conveys this, and clearly indicates that they are looking at personal action on the Nexus:

> The "nexus" is where food, water and energy systems intersect. It takes water (lots of it) to create food and energy. It also takes energy to move, heat and treat water and to produce food, and sometimes we even use food crops as a source of energy. Find out how anyone – from a single person to an entire nation – can make more sustainable choices with nexus thinking.
>
> (GRACE, 2014).

Just before the 2014 Nexus conference GRACE released its 'Meet the Nexus' guide, which breaks down the Nexus concept into easy-to-digest parts by revealing the hidden connections between food, water and energy in grocery store aisles, at home and in the kitchen.

> "When I tell people that it takes 42 gallons of water to make one slice of pizza, they are surprised," says Kyle Rabin, Director of Programs at GRACE. "We hope this guide leads people to consider the profound effects of our everyday food, water and energy decisions. Readers may never again look at a cheese pizza without recognizing the precious resources that go into it. At the very least, we hope they will think twice before tossing out their leftovers."
>
> (GRACE, 2014)

In 'Meet the Nexus' the Food, Water and Energy Nexus is presented as simply the intersection of systems:

- It takes water and energy to produce the food you eat.
- Energy is used to move water to your home, to heat that water, and then later to clean up the water that flows down the drain.
- Water is required to run power plants safely and to produce oil, gas and coal.
- Some food crops are turned into fuel for vehicles.

"Our report breaks this complex idea down to nine simple tips that illustrate how making even one good decision about food, water or energy resources can have a positive impact on the others," says Peter Hanlon, the guide's lead author. "For example, switching off a light doesn't just save electricity because conventional power plants require a lot of water, and so does the production of fuel used at those plants. Consider it more bang for the sustainable-behavior buck."

(Hanlon, 2014)

Next steps

The SDG negotiation process resumed in January and culminated in the UN Special Summit in September 2015, which gave a clear direction to the international community. It delivered 'a' transformative agenda, but perhaps not 'the' transformative agenda. In the short term up to March 2016 the Nexus community needs to come forward with a suggested set of indicators for the relevant targets to be included in the 100+ global indicators, which governments will agree to. In the longer term, once the goals, targets and indicators are set it will all be about developing the toolkits and frameworks to enable policy makers, stakeholders and intergovernmental bodies to change the way that these sectors have been approached. The bi-annual meetings of a new World Forum on Sustainable Development Data under the UN Statistical Commission will be a very important area to promote Nexus indicators. In addition to the 100+ global indicators they will set non-core indicators through a 'complementary architecture'. This could possibly be developed through 'thematic catalogues', which would allow the collection of other indicators already being used at the local, sub-national, national and international levels. This could lead to more citizen data collection in some cases.

Final thoughts

Sector conversations have evolved dramatically since the 1992 Earth Summit. The UN CSD, set up to review the implementation of Agenda 21 and the JPoI, had some success in moving forward a number of single-sector issues such as chemicals, forests and oceans, but failed to deal effectively with the interlinkages.

The Nexus is about the inter-connectedness of water, food and energy resources and their better management through enhanced governance across sectors and scales.

This approach has much promise. It is developing a community of actors who are assembling connections between case studies that show how to deal with trade-offs, and examples to inform the real-world decisions that governments, intergovernmental bodies and stakeholders make daily.

The number of Nexus initiatives that have started since the Rio+20 Conference, is accelerating, points towards increasing recognition, acceptance and engagement. Some of these initiatives are reflected in chapters in this book, such as the work of FE2W and the work on the urban dimension of the Nexus.

We have a short time available in which to choose a path that will enable us all to live on this planet more sustainably, more justly and more equitably. That future is far from guaranteed, and the wise warning of Martin Luther King should keep us focused:

> Over the bleached bones of numerous civilizations are written the pathetic words: "Too late!"

References

Berkowitz, D. (2001) *Does Privatization Enhance or Deter Small Enterprise Formation?* Pittsburgh: University of Pittsburgh. Available online at: www.pitt.edu/~dmberk/bhfinal3.pdf

Bird, J. (2014) *Foreword to Setting and Achieving Water Related Sustainable Development Goals, IWMI.* Available online at: www.iwmi.cgiar.org/Publications/Books/PDF/setting_and_achieving_water-related_sustainable_development_goals.pdf

Bird, J., Dodds, F., McCormick, P. and Shah, T. (2014) *Water-Food-Energy Nexus in Setting and Achieving Water Related Sustainable Development Goals, IWMI.* Available online at: www.iwmi.cgiar.org/Publications/Books/PDF/setting_and_achieving_water-related_sustainable_development_goals.pdf

Clean Tech (2014) *Texas is Fracked: More than 30 Towns Will be Out of Water due to Fracking.* Available online at: http://cleantechnica.com/2013/08/18/30-texas-towns-will-be-out-of-water-because-fracking/

Declaration on Renewable Energy (2002) *The Way Forward on Renewable Energy*, signed by Bulgaria, Cyprus, Czech Republic, Estonia, the European Union, Hungary, Iceland, Latvia, Lithuania, Malta, New Zealand, Norway, Poland, Romania, Slovakia, Slovenia, the Alliance of Small Island States, Switzerland and Turkey. Available online at: http://www.un.org/events/wssd/statements/joint-declaration.htm

Dublin Principles (1991) *Dublin Statement on Water and Sustainable Development.* Available online at: http://en.wikipedia.org/wiki/Dublin_Statement

Eid, U. (2014) *Urban NEXUS – Expert Statements.* Available online at: www.iclei.org/urbannexus.html

Food and Agriculture Organization (FAO) (1996) The Rome Declaration on World Food Security. Available online at: www.fao.org/wfs/index_en.htm

FE2W (2014) FE2W website. Available online at: http://www.fe2wnetwork.org/

Germany (2001) *Water a Key to Sustainable Development.* Available online at: http://www.un.org/esa/sustdev/sdissues/water/BonnConferenceReport.pdf

Germany (2004) *Bonn Renewable Conference Outcomes.* Available online at: http://en.wikipedia.org/wiki/International_Renewable_Energy_Conference

GIZ and ICLEI (2014) *Operationalizing the Urban Nexus: Towards resource-efficient and integrated cities and metropolitan regions.* Available online at: http://www2.giz.de/wbf/4tDx9kw63gma/02_UrbanNEXUS_CaseStudy_Nashik.pdf

Global Environmental Facility (1991) *GEF History.* Available online at: www.thegef.org/gef/whatisgef

GRACE (2014) Front page of the GRACE Communications Foundation website. Available online at: www.gracelinks.org/468/nexus-food-water-and-energy

Hanlon, P. (2014) GRACE Communications Foundation website. Available online at: www.gracelinks.org/media/pdf/nexusguide_press percent20release_final.pdf

International Fund for Agricultural Development (2014) *Water facts and figures.* Available online at: www.ifad.org/english/water/key.htm

IPCC (1990) *Policymakers' Summary prepared by IPCC Working Group 1*. Available online at: www.ipcc.ch/ipccreports/far/wg_I/ipcc_far_wg_I_spm.pdf

IPCC (2008) *Technical Paper VI on Climate Change and Water*. Available online at: www.ipcc.ch/pdf/technical-papers/climate-change-water-en.pdf

Lula, S. (2007) *Speech to the United Nations General Assembly*. Available online at: www.un.org/apps/news/story.asp?NewsID=23952&Cr=general&Cr1=debate#.VG1NyvnF98E

Mbeki, T. (2006) Speech to the United Nations General Assembly. Available online at: www.dfa.gov.za/docs/speeches/2006/mbek0919.htm

Meadows, D., Meadows, G., Randers, J. and Behrens III, W. W. (1972) *The Limits to Growth*. New York: Universe Books. ISBN 0-87663-165-0. Available online at: www.donellameadows.org/wp-content/userfiles/Limits-to-Growth-digital-scan-version.pdf Melbourne, Australia: Melbourne Sustainable Society Institute, The University of Melbourne. p. 16. ISBN 978-0-7340-4940-7.

Organization of Economic Cooperation and Development (2014) *Official Development Assistance 2013*. Available online at: www.oecd.org/newsroom/aid-to-developing-countries-rebounds-in-2013-to-reach-an-all-time-high.htm

Pepsico (2014) *From the Pepsico website: Water Stewardship*. Available online at: http://www.pepsico.com/Purpose/Environmental-Sustainability/Water

SAB Miller and WWF International (2009) *Identifying and Addressing Water Risks in the Value Chain*. Available online at: www.sab.co.za/sablimited/action/media/downloadFile?media_fileid=918

Sunday Independent (1999) *The Hidden Tentacles of the World's Most Secret Body*. Available online at: http://en.wikipedia.org/wiki/1999_Seattle_WTO_protests

Telegraph (2014) *Martin Dent Obituary*. Available online at: http://www.telegraph.co.uk/news/obituaries/10875464/Martin-Dent-obituary.html

Turner, G. (2008) *A Comparison of the Limits to Growth with Thirty Years of Reality*. Socio-Economics and the Environment in Discussion (SEED). CSIRO Working Paper Series Number 2008-09. June 2008. ISSN 1834-5638. 52 pp. Available online at: www.csiro.au/Outcomes/Environment/Population-Sustainability/SEEDPaper19.aspx

Turner, G. (2014) In L. Rickards (ed.) *Is Global Collapse Imminent?* MSSI Research Paper No. 4. Available online at: www.sustainable.unimelb.edu.au

United Nations Commission on Transnational Corporations (1991) *Transnational Corporations and Sustainable Development* (Chapter 41, Agenda 21). Available online at: http://unctc.unctad.org/data/ec1019922a.pdf

United Nations (1970) *International Strategy for the Second United Nations Development Decade*. Available online at: www.unmillenniumproject.org/documents/overviewEng55-65LowRes.pdf

United Nations (1987) *Our Common Future*, New York: UN.

United Nations (1992) *Agenda 21*. Available online at: http://sustainabledevelopment.un.org/content/documents/Agenda21.pdf

United Nations (1997) *The Program for the Further Implementation of Agenda 21*. Available online at: www.un.org/documents/ga/res/spec/aress19-2.htm

United Nations (2002a) Committee on Economic, Social and Cultural Rights adopted General Comment No. 15 on the Right to Water. Available online at: www1.umn.edu/humanrts/gencomm/escgencom15.htm

United Nations (2002b) Johannesburg Plan of Implementation. Available online at: www.un.org/esa/sustdev/documents/WSSD_POI_PD/English/WSSD_PlanImpl.pdf

United Nations (2010) *General Assembly Resolution 64/292: The human right to water and sanitation*. Available online at: www.un.org/es/comun/docs/?symbol=A/RES/64/292&lang=E

United Nations (2012) *The Future We Want*, New York: UN.

United Nations (2014) *Report of the Open Working Group on Sustainable Development Goals established pursuant to General Assembly resolution 66/288*. Available online at: www.un.org/en/ga/search/view_doc.asp?symbol=A/RES/68/309

UN Water (2014) Home page of UN Water. Available online at: www.unwater.org/about/en/

World Commission on Environment and Development (1987) *Our Common Future*. Oxford: Oxford University Press. p. 27. ISBN 019282080X. Available online at: www.un-documents.net/ocf-ov.htm#1.2

World Resources Institute (2002) Press Release at the end of WSSD. Available online at: http://en.wikipedia.org/wiki/Type_II_Partnerships#cite_note-WRI-8

World Water Commission (2000) *World Water Vision*. Available online at: www.worldwatercouncil.org/index.php?id=961&L=0t

2 Sustainable Development Goals and policy integration in the Nexus

David Le Blanc

Introduction

In 2014, after more than a year of intergovernmental work, UN Member States proposed a set of Sustainable Development Goals or SDGs (United Nations, 2014a). The SDGs will succeed the Millennium Development Goals (MDGs) as reference goals for the international community for the period 2015–2030. The development of the new set of goals is widely seen as an ambitious agenda, as these goals cover a much broader range of issues than their predecessors, they aim to be universal – that is, applicable to all countries and not only developing countries – and they have to serve as guideposts for a difficult transition to sustainable development, which has eluded the international community since the Earth Summit in 1992.

All actors familiar with the Nexus, from practitioners to policy-makers, have long recognized the need for policy integration across the sectors covered by the Nexus. This need comes from the multiple links, trade-offs and synergies that exist among potential objectives and targets in the Nexus (Bazilian *et al.*, 2011). In that context, and given the tendency of policy to operate in silos, a conducive framework for policy integration has tremendous value (Weitz *et al.*, 2014).

Internationally agreed development goals and targets carry considerable weight in development circles, both in terms of putting some issues at the forefront of political attention, and because they often provide the basic common benchmark against which the institutions tasked with policy-making, at the international level but also to some degree at the national level, operate. For development institutions that structure their work around internationally agreed goals in particular, the new goals can be expected to provide a framework around which policy and action will be justified and organized. The more goals and targets refer to one another, the easier integration arguably is, as development agencies concerned with a specific goal (e.g. energy or water) will have to take into account targets that refer to other goals in designing, implementing and monitoring their work. Thus, provided that they are built to foster integration (for example, because targets under specific goals also refer to other goals), internationally agreed goals can serve as powerful allies in the drive for policy integration, in the Nexus as well as in other policy clusters.

This is why experts in the Nexus community have been looking to the proposed SDG framework with questions about how well the SDGs would facilitate considerations of policy integration. As with any new approach, there are many ways to look at the new framework in terms of how well it meets expectations from specific groups. One way is to ask whether it does better in this regard than what existed previously (the MDGs). Another is to ask whether it could have done better. This is a version of the glass half-full or half-empty story.

This chapter does not seek to answer the two questions above. Rather, it tries to shed light on the question of whether the SDGs can provide a conducive framework for policy integration in the Nexus. To do so, I assess the degree to which the structure of the SDGs takes note of existing links among targets within the Nexus, and compare it to what scientists say about existing links.

The remainder of the chapter is organized as follows. The next section briefly reviews what is known about interactions in the Nexus. The third section documents how the SDG and targets link areas within the Nexus. In the fourth section I contrast the links that are made within the SDGs themselves and the links that have been documented by scientists and practitioners in the Nexus. The final section concludes.

Interactions in the Nexus

Before going any further, we have to define the contours of the Nexus that we will use in this chapter. The Nexus comes in different hues, and different communities adopt different perimeters for it and give it different names (UN, 2014b). Energy analysts typically refer to climate–land–energy–water (CLEW) strategies or energy–food–water strategies, whereas water analysts tend to refer to the "nexus" and in particular the Water–Energy–Food Security Nexus (WEF).[1] The CLEW community has brought analytical and practical clarity to the policy importance of links among the Nexus sectors through the use of planning and modelling tools (Rogner *et al.*, 2010; Bazilian *et al.*, 2011; Howells *et al.*, 2013). The global model for CLEW created by Weirich (2013) adds to these four sector linkages with materials consumption. Whether climate is formally included in the Nexus or not, climate change has strong connections with the three components of the Nexus. In the remainder of this chapter, I consider the CLEW Nexus.

Interactions among sectors and sub-sectors in the CLEW Nexus have been studied both in theory and through dedicated modelling at the global level (Hermann *et al.*, 2011; Hoff, 2011; Rogner, 2010; Bazilian *et al.*, 2011; Hermann *et al.*, 2012; Weirich, 2013; Howells *et al.*, 2013), the regional level (Swierinski, 2012; UNECE, 2013), as well as the national and sub-national levels (Hermann *et al.*, 2012; Karlberg *et al.*, 2015; Morrison, 2012; IRENA, 2013; Welsch *et al.*, 2014).[2]

The range of existing studies all highlight the need to consider interactions, synergies and trade-offs among sectors. For illustrative purposes, Table 2.1, taken from the *Prototype Global Sustainable Development Report* published in 2014 by the United Nations, lists important inter-linkages in the CLEW Nexus. Another illustration of the links is given by Weitz *et al.* (2014, p. 46), who document links across targets belonging to the Nexus goal areas. An extensive list of relevant inter-linkages has been assembled by the US Pacific Northwest National Laboratory (Skaggs *et al.*, 2012).

Another lesson from the range of available studies is that the nature and importance of the links among Nexus areas differ markedly according to location and scale (Howells *et al.*, 2013; Weitz *et al.*, 2014; UN, 2014a). In some locations the links between energy and water may be the most important from a policy perspective. In other locations, other links may be more important.

Table 2.1 Selected inter-linkages between climate, land/food, energy and water. Source: UN (2014a), adapted from Weirich (2013), Rogner (2010), Hoff (2011) and Howells *et al.* (2013).

Impacts of the issues listed below on those listed on top	*Climate*	*Land/Food*	*Energy*	*Water*
Climate		Climate change and extreme weather affect crop productivity and increase water demand in most cases.	Climate change alters energy needs for cooling and heating and impacts the hydropower potential.	Climate change alters water availability and the frequency of droughts and floods.
Land/Food	Greenhouse gas emissions from land use change (vegetation and "soil carbon") and fertiliser production.		Energy is needed for water pumping, fertiliser and pesticide production, agricultural machinery and food transport.	Increased water demand due to intensification of agriculture, and effects on the N/P cycles.
Energy	Fuel combustion leads to GHG emissions and air pollution.	Land use for biofuels and renewable energy tech. (solar, wind, hydro, ocean), crop/oil price correlation.		Changes in river flow, evaporation in hydropower dams, biofuels crop irrigation, fossil fuel extraction (esp. unconventional).
Water	Changes in hydrological cycles affect local climates.	Changes in water availability for agriculture and growing competition for it affect food production.	Water availability for biofuels, energy use for desalination but also storage of renewable energy as fresh water.	

What do the SDGs say about the links in the CLEW Nexus?

The set of SDGs that was put forward by the Open Working Group is the result of intergovernmental discussions. As such, it constitutes a normative piece, which frames global goals and targets that the international community sets for it. As a compromise reflecting a multiplicity of concerns and interests, the set of SDGs taken as a whole is not based on any particular interpretation of the world, and nor does it reflect a specific, coherent systemic view of how the socio-economic engine works and delivers outcomes along all the dimensions covered by the goals.

At a first level, compared to the MDGs, the SDGs promise to give greater visibility to a variety of sector clusters that are important for policy because they encompass a much broader range of areas. This is particularly true for the CLEW Nexus, as each of its components was dedicated to a goal under the SDGs. Agriculture and food (goal 2), water (goal 6), energy (goal 7) and climate change (goal 13) are thus well identified in the new framework, and this should contribute to increasing the visibility of the Nexus.

At a second level, scientists and practitioners alike have noted that it is important that the goals and targets should reflect strong interconnections among goal areas from the biophysical and socio-economic points of view (Griggs *et al.*; 2014, Weitz *et al.*, 2014; ICSU and ISSC, 2015). As argued above, the more the SDGs and targets link to one another, the easier policy integration becomes. To what extent do the SDGs achieve this for the CLEW Nexus? Based on the wording of the targets in the SDGs, the answer is, very little. Table 2.2 shows the targets under goals 2, 6, 7, 13 and 15 that explicitly refer to other Nexus areas.

Table 2.2 Explicit references to other Nexus areas in the Nexus SDGs. Source: Author's elaboration.

Goals in the Nexus	*Targets referring to other Nexus goals*
Goal 2: Food and agriculture	Target 2.4 refers to climate change and terrestrial ecosystems: "by 2030 ensure sustainable food production systems and implement resilient agricultural practices that increase productivity and production, that help maintain ecosystems, that strengthen capacity for adaptation to climate change, extreme weather, drought, flooding and other disasters, and that progressively improve land and soil quality"
Goal 6: Water	Target 6.6 refers to terrestrial ecosystems: "by 2020 protect and restore water-related ecosystems, including mountains, forests, wetlands, rivers, aquifers and lakes"
Goal 7: Energy	No target explicitly refers to the others in the Nexus
Goal 13: Climate change	No target explicitly refers to the others in the Nexus
Goal 15: Terrestrial ecosystems	Target 15.3 refers to food and agriculture: "by 2020, combat desertification, and restore degraded land and soil, including land affected by desertification, drought and floods, and strive to achieve a land-degradation neutral world"

Thus, it is clear that the political framework that the SDGs provide does not reflect the multiplicity of links that have been identified in the Nexus literature. This is in contrast with other parts of the SDG system, which show more connections among goal areas. Indeed, the proposed goals and targets can be seen as a network, in which links among goals exist through targets that explicitly refer to multiple goals. This approach highlights where the political process that created the SDGs made links between goals. The resulting network and mapping, which reflect the results of negotiations in an intergovernmental context, can be thought of as a "political mapping" of the sustainable development universe, as opposed to scientific mappings based on natural and social science insights. A look at the whole system based on this method shows that some parts of the system are quite well connected (Le Blanc, 2015). Figure 2.3 (below) shows the resultant mapping. It is clear that the links among energy, water, food and agriculture and climate change are few.[3]

Hence, at this second level SDGs will be of limited use in providing guidance to address the various links that exist in the Nexus. To some extent, this should not come as a surprise. The SDGs, as political constructions for which one parameter was that the goals should be "limited in number", could not possibly address all the relevant links among goal areas – there are simply too many of them. Moreover, the fact that the importance and relevance of the links vary across location and scale, as mentioned in the introduction, means that the presence in the SDGs of many targets linking the Nexus areas would be of limited use for concrete policy-making in many cases.

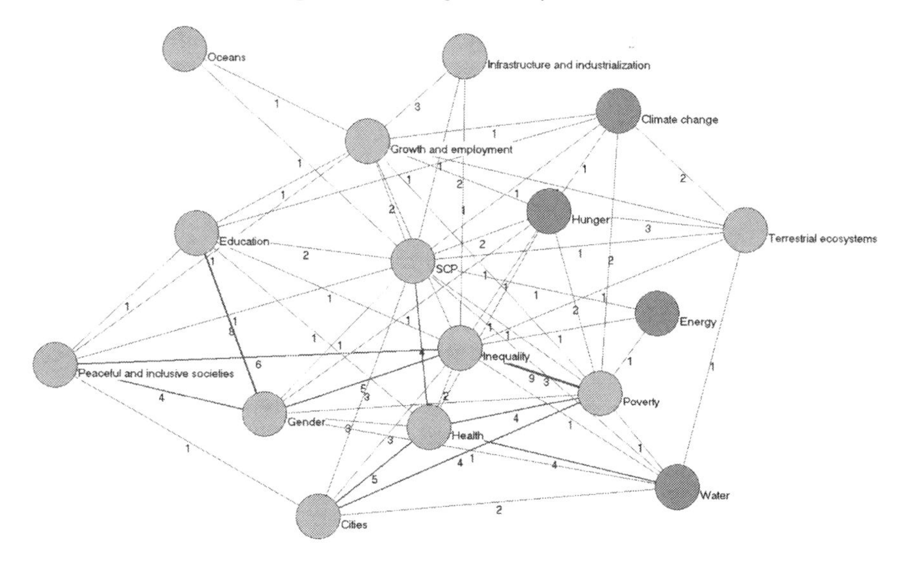

Figure 2.1 Explicit links made by targets among SDG areas: the whole SDG system. Source: Le Blanc (2015). Note: the numbers on the map indicate the number of targets linking the goals. For example, SDG 16 on peaceful and inclusive societies is linked with SDG 5 on gender through four targets.

What do scientists say about the links among targets in the Nexus goals?

The "political" mapping just explored above can be contrasted with what scientists say about the links in the Nexus. Several approaches can inform us here. The first is the one presented in section 1, which does not consider the SDGs as such but simply focuses on biophysical, social and economic links among the Nexus areas (Bazilian *et al.*, 2011; Welsch *et al.*, 2014; Skaggs *et al.*, 2012; and other references in section 1). The number of links considered by planning and modelling tools that have been applied to CLEW tends to be high. Overall, the sheer number of interactions among the Nexus areas mentioned in these studies is enough to show that the SDG targets do not capture most of these interactions.

A second approach is to start from the SDGs and consider the links between SDG targets in terms of one goal and all the other goals. This is the approach adopted in a recent paper by the International Council for Science (ICSU and ISSC, 2015). As this approach is similar to the one which underlies the "political" mapping, presented above, it allows for a straightforward comparison with the latter. ICSU and ISSC (2015) asked small groups of expert scientists on each goal in the SDGs to, inter alia, mention the links between the goal being considered and targets under the other goals. The results, consolidated by goals, are shown in Figure 2.2 below. As can be seen from Figure 2.2, the resulting network is much more densely connected than the one shown in Figure 2.1. Each of the goals links to the majority of targets under the other goals – and to the majority of targets under the non-Nexus goals as well.[4] This clearly shows that there are many more connections among goals than are reflected explicitly in the SDGs.

If we focus on the CLEW Nexus itself, the ICSU–ISCC paper suggests links between most of the targets under goals 2, 6, 7 and 13 to other goals in the Nexus. The resulting links among targets in the CLEW network are shown in Figure 2.3. Some of the targets are actually assessed as linking the four goals; many of them link three of the four goals. This stands in sharp contrast with the limited number of links appearing in Figure 2.1. Lastly, Figure 2.4 shows the links between the CLEW Nexus considered as a whole and other goals, based on the same data. From the perspective of policy integration this picture is instructive, as it reminds us that actions in the CLEW Nexus may potentially impact on all other goals.

In a third and slightly different approach, Weitz *et al.* (2014, p. 46) consider links across targets belonging to the Nexus goal areas and distinguish three categories of links: targets that reinforce each other; targets that are dependent on each other; and targets that impose conditions on each other.

Overall, the conclusion is clear: from an operational policy perspective there are many links among Nexus goals, and most of those are not reflected in the SDGs.

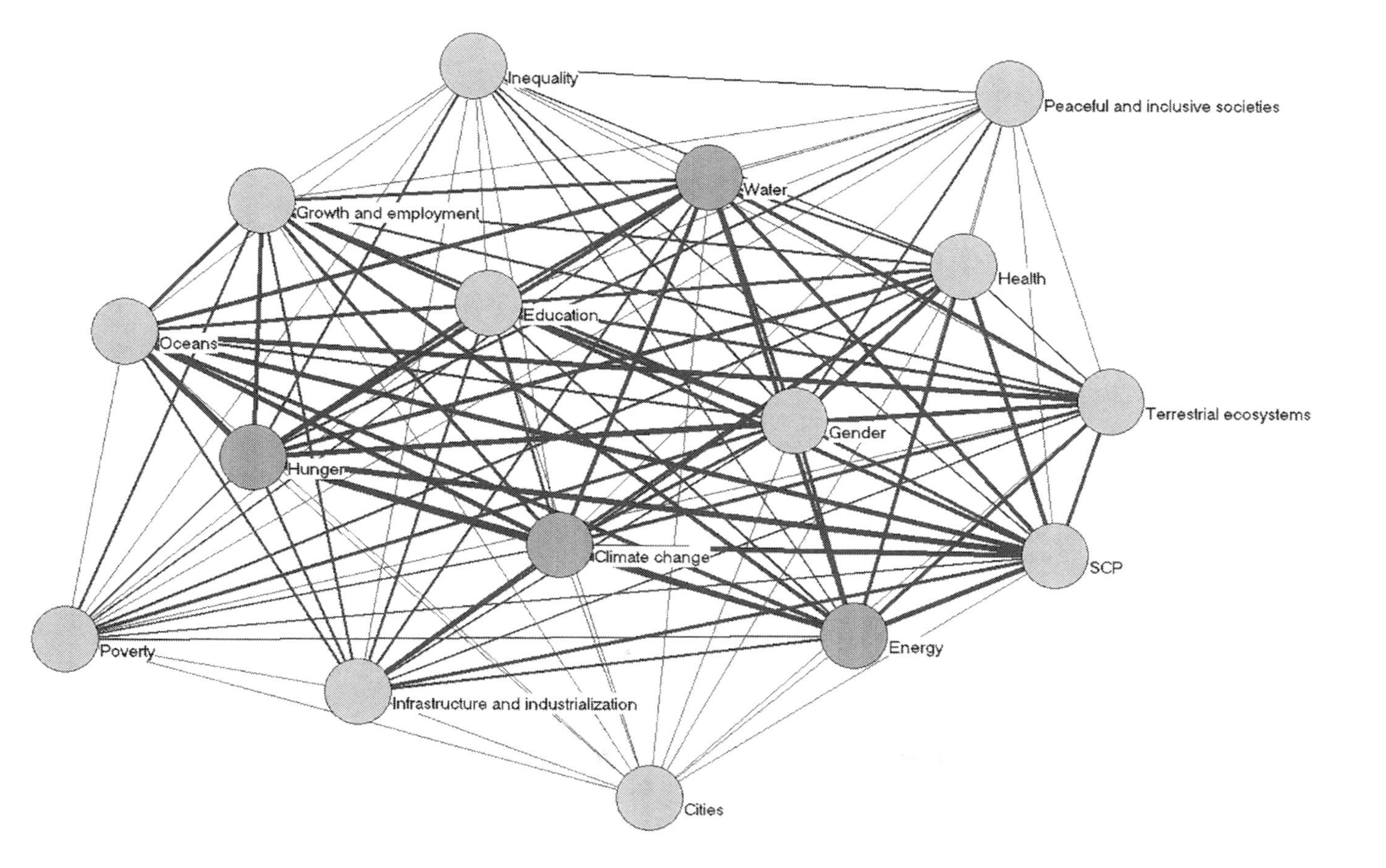

Figure 2.2 Links between SDG targets and other goals made by scientists in a review of SDGs: The whole SDG system. Source: Le Blanc (2015). Note: In the ICSU–ISSC paper, the chapter on goal 11 (cities) does not record links with targets under other goals. This explains the lower number of links starting from that goal than among many other goals.

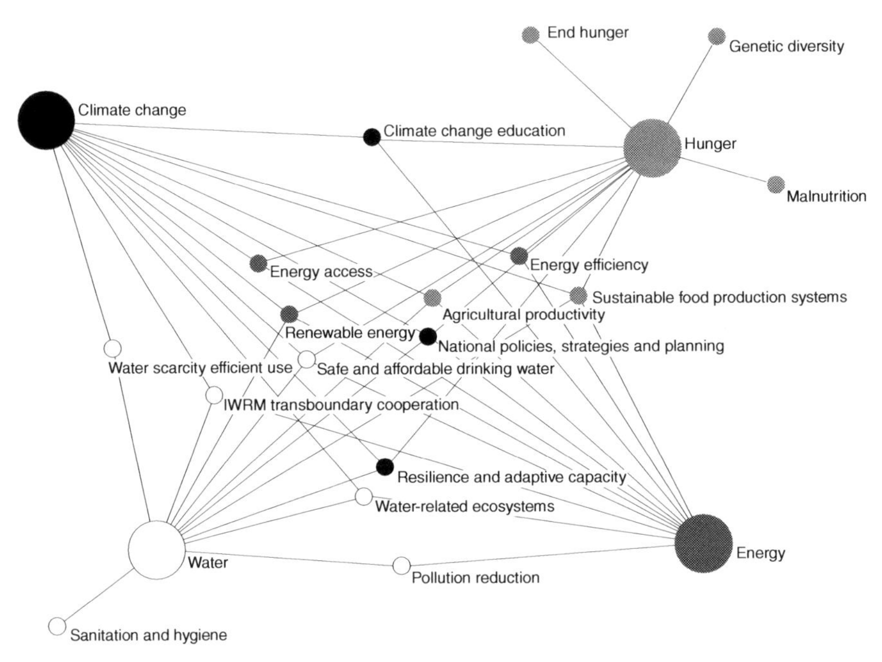

Figure 2.3 Links among targets in the CLEW Nexus put forward by the ICSU–ISSC scientific review of SDGs. Source: Le Blanc (2015). Note: Broader circles indicate goals. Smaller circles indicate targets. Targets are of the same shading as the goal to which they belong.

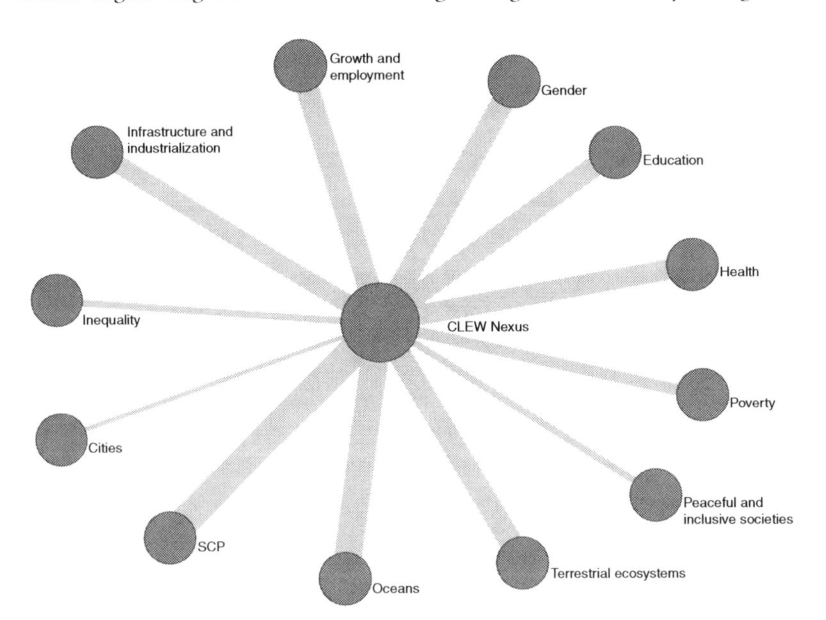

Figure 2.4 Strength of the links among the CLEW Nexus and other goals, as put forward by the ICSU–ISSC scientific review of SDGs. Source: Author's elaboration, from data in ICSU–ISSC (2015). Note: In the ICSU–ISSC paper, the chapter on goal 11 (cities) does not record links with targets under other goals. This explains the lower number of links starting from that goal than among many other goals.

Conclusion

In conclusion, the Nexus will likely benefit from increased visibility in the policy arena due to the fact that food, water, energy and climate change all have dedicated goals in the proposed SDGs. Some of the targets under the other goals also refer to one or several of the Nexus areas, which should facilitate the consideration of policy links between the Nexus and the rest of the SDG areas.

However, the SDGs targets weakly capture the interactions, synergies and trade-offs among the Nexus areas. This is also true for the links that exist among CLEW areas and other SDG areas. Hence, SDGs will probably be of limited use in providing guidance to address the various links that exist in the Nexus. Operationalization of these links in policy will need to be done by other means. To some extent, this would have needed to be the case anyway, given the wide variability in the importance and relevance of different links across location and scale, which calls for studies, modelling and policy at the appropriate scale and for policies that are, to some degree, idiosyncratic. Many concrete approaches have been developed in recent years to visualize critical links, trade-offs and synergies in the Nexus and translate them into policies. Given the importance of the Nexus, they will undoubtedly prove useful in years to come.

DISCLAIMER: The views expressed herein are those of the author and do not necessarily reflect the views of the United Nations.

Notes

1 Analysts with a food security perspective as a starting point have used a combination of the above (UN, 2014b, ch. 6).
2 For more references to regional and national case studies, see UN (2014b).
3 In this approach, links between targets under one goal and other goals are created only if the wording of a target explicitly refers to these goals. The resulting network is then consolidated by goal.
4 Because links were recorded by different teams for each goal, the ICSU–ISCC paper may suffer from lack of consistency, in the sense that some teams have been more "generous" in linking a goal with other goals. Also, the linking exercise was not done in a similar fashion for goal 11 (cities).

References

Bazilian, M., Rogner, H.-H., Howells, M., Hermann, S., Arent, D., Gielen, D., Steduto, P., Mueller, A., Komor, P., Tol, R.S.J. and Yumkella, K.K. (2011) 'Considering the energy, water and food nexus: Toward an integrated modeling approach.' *Energy Policy*, 39(12), December, 7896–7906. Available online at: www.sciencedirect.com/science/article/pii/S0301421511007282

Griggs, D., Stafford Smith, M., Rockström, J., Öhman, M.C., Gaffney, O., Glaser, G., Kanie, N., Noble, I., Steffen, W. and Shyamsundar, P. (2014) 'An integrated framework for sustainable development goals', *Ecology and Society*, 19(4), 49. http://dx.doi.org/10.5751/ES-07082-190449.

Hermann, S., Rogner, H.-H., Howells, M., Young, C., Fischer, G., & Welsch, M. (2011) *In The CLEW Model – Developing an integrated tool for modelling the interrelated effects of Climate, Land use, Energy, and Water (CLEW)*. 6th Dubrovnik Conference on Sustainable Development of Energy, Water and Environment Systems – Proceedings. Stockholm, Sweden.

Hermann, S., Welsch, M., Segerstrom, R., Howells, M., Young, C., Alfstad, T., Rogner, H.-H. and Steduto, P. (2012) 'Climate, land, energy and water (CLEW) interlinkages in Burkina Faso: An analysis of agricultural intensification and bioenergy production', *Natural Resources Forum*, 36, 245–262.

Hoff, H. (2011) *Understanding the Nexus. Background Paper for the Bonn 2011 Conference: The Water, Energy and Food Security Nexus*. Stockholm, Sweden: Stockholm Environment Institute.

Howells, M., Hermann, S., Welsch, M., Bazilian, M., Segerstrom, R., Alfstad, T., Gielen, D., Rogner, H., Fischer, G., Velthuizen, H. van, Wiberg, D., Young, C., Roehrl, R., Mueller, A., Steduto, P. and Ramma, I. (2013) 'Integrated analysis of climate change, land use, energy and water strategies.' *Nature Climate Change*, 3, July 2013.

ICSU and ISSC (2015) *Review of Targets for the Sustainable Development Goals: The Science Perspective*, Paris, February.

Intergovernmental Panel on Climate Change (2014) *Climate Change 2014: Mitigation of Climate Change*. Contribution of Working Group III to the Fifth Assessment Report of the Intergovernmental Panel on Climate Change, Cambridge University Press, Cambridge, United Kingdom and New York, NY, USA.

IRENA (2013) *Renewable Energy Policy in Cities: Selected Case Studies*, Abu Dhabi: IRENA. Available online at: http://www.irena.org/menu/index.aspx?mnu=Subcat&PriMenuID=36&CatID=141&SubcatID=286

Karlberg, L., Hoff, H., Amsalu, T., Andersson, K., Binnington, T., Flores-López, F., de Bruin, A., Gebrehiwot, S.G., Gedif, B., zur Heide, F., Johnson, O., Osbeck, M. and Young, C. (2015) 'Tackling complexity: Understanding the food-energy-environment nexus in Ethiopia's Lake Tana Sub-basin.' *Water Alternatives*, 8(1), 710–734.

Le Blanc, D. (2015) *Towards integration at last? The Sustainable Development Goals as a network of targets*. DESA Working paper No 141, UN Department of Economic and Social Affairs, New York, March.

Morrison, R. (2012) *Sustainable Development in Jamaica through biofuels with special focus on Climate-Land-Energy-Water (CLEW)*. European Joint Masters in management and engineering of the environment and energy, KTH Royal Institute of Technology, Sweden, and Queens University Belfast.

Rogner, H.-H. (2010) APPENDIX to *CLEWS: Climate, Land, Energy, Water Strategies, A Pilot Case Study*, Vienna, Austria.

Skaggs, R., Hibbard, K., Janetos, T. and Rice, J. (2012) *Climate and Energy-Water-Land System Interactions*, Technical Report to the US Department of Energy in Support of the National Climate Assessment. Pacific Northwest National Laboratory, PNNL-21185, Richland, WA, USA.

Swierinski, A. (2012) *The Water-Land-energy Nexus in the Pacific – a case study*. IRENA.

United Nations (2014a) *General Assembly, Report of the Open Working Group of the General Assembly on Sustainable Development Goals*, A/68/970, August.

United Nations (2014b) *Prototype Global Sustainable Development Report*, Division for Sustainable Development, New York, June, Available online at: http://sustainabledevelopment.un.org/content/documents/1454Prototype%20Global%20SD%20Report.pdf

UNECE (2013) *A proposed approach to assessing the Water-Food-Energy-Ecosystems Nexus under the UNECE Water Convention*. Discussion paper prepared by the UNECE secretariat with input from Finland, the Food and Agriculture Organization of the United Nations (FAO), the Stockholm Environment Institute (SEI) and the Stockholm International Water Institute (SIWI), 4 April, 2013.

Wang, J. (2012) 'China's water-energy nexus: Greenhouse-gas emissions from groundwater use for agriculture', *Environmental Research Letters*, 7(1).

Weirich, M. (2013) *Global resource modelling of the Climate, Land, Energy and Water (CLEWS) Nexus using the open source energy modelling system (OSEMOSYS)*. Internship report, available online at: www.diva-portal.org/smash/get/diva2:656757/FULLTEXT01.pdf

Weitz N., Nilsson, M. and Davis, M. (2014) 'A nexus approach to the post-2015 agenda: Formulating integrated water, energy and food SDGs.' *SAIS Review of International Affairs*, 34, 37–50.

Welsch, M., Hermann, S., Howells, M., Rogner, H.-H., Young, C., Rammad, I., Bazilian, M., Fischer, G., Alfstad, T., Gielen, D., Le Blanc, D., Röhrl, A., Steduto, P. and Müller A. (2014) 'Adding value with CLEWS – Modelling the energy system and its interdependencies for Mauritius', *Applied Energy*, 113, 1434–1445.

3 Nexus scientific research

Theory and approach serving sustainable development

Joachim von Braun and Alisher Mirzabaev

Introduction

Agricultural, water, energy, industrial and climate policies influence each other and jointly determine outcomes for people and the environment, creating complex trade-offs and potential synergies (Ringler *et al.*, 2013). Seeking to address these trade-offs from the point of view of a particular sector alone would create intricate feedback loops, potentially leading to unintended consequences and negative externalities. The Water–Energy–Food Security (WEF) Nexus is a conceptual framework that recognizes such interconnections among these broad domains and seeks to develop joint solutions that mitigate the trade-offs and promote synergies among them (Hoff, 2011).

The WEF Nexus is not only about sectoral policy practices, but also about how scientific research is conducted on these Nexus domains. In this regard, Nexus thinking is not an entirely new phenomenon in science. It closely relates to systems theory, which advocates that scientific research needs to consider the whole environmental and economic system, including various interactions between the parts of the system, and not limit itself to the specific constituent parts of the whole in isolation from the rest of the system (von Bertalanffy, 1956).

Nexus thinking in science is manifested through such concepts as multidisciplinarity, interdisciplinarity, and trans-disciplinarity. Multidisciplinary approaches seek to combine insights from different disciplines, where each discipline follows its own set of rules and methods and is not necessarily influenced by other disciplines in its analytical processes. On the other hand, interdisciplinary research seeks to integrate different disciplines more closely through collaboration and mutual cross-fertilization of concepts, methods and data (Miller, 1982), i.e. crossing the boundaries between the disciplines (Mollinga, 2010). Finally, trans-disciplinary research goes a step further than interdisciplinary research by also embracing the inputs and ideas from relevant non-scientific stakeholders into the research agenda, treating them not solely as objects of the research, but as active participants in the process. Figure 3.1 shows the occurrence of these three concepts, as well as of the concept of disciplinarity in the literature since the beginning of the nineteenth century.

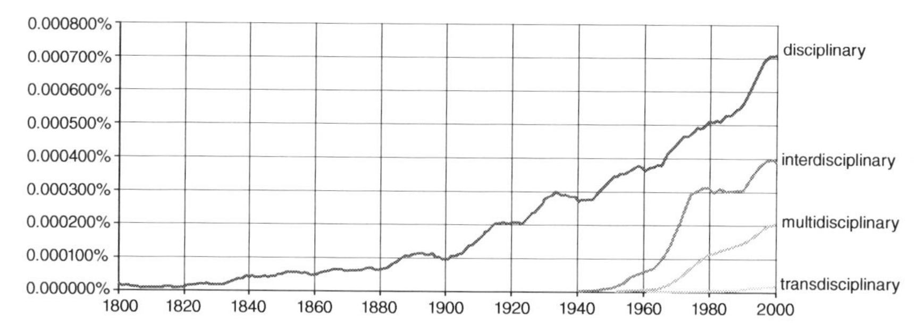

Figure 3.1 Occurrence of terms describing the types of research by their disciplinarity.

Figure 3.1 shows that the concept of disciplinarity in science was already present in the early 1800s, but the concepts of interdisciplinary and multidisciplinary research emerged more widely in the literature only after the Second World War, while the concept of trans-disciplinary research is a relatively recent term, only making an appearance in the early 1980s.[1]

In this regard, Nexus science is by definition trans-disciplinary, since it brings together not only different disciplines but also different stakeholders, including non-scientific ones, in the societal and political process of decision making, with the purpose of harnessing these diverse sources of knowledge to find more accurate solutions for existing problems of sustainable development.

The current chapter first explores the factors of global change that are generating an ever greater demand for Nexus approaches in scientific research. Second, it presents the key conceptual underpinnings of Nexus science. Next, the chapter provides some practical examples of modelling the Nexus domains (or sectors) at different scales, and shows how previous modelling approaches need to be enhanced to better suit the Nexus concepts. Finally, the chapter concludes with suggestions for ways to integrate Nexus thinking into science management and organization, including the implications of investment strategies for research on sustainable development across the Nexus domains and Nexus governance, taking into account the realities of the Nexus political economy.

Background

A confluence of several factors is bringing about tremendous shifts in how research on sustainable development issues needs to be conducted. These changes make earlier approaches based on compartmentalization of scientific research suboptimal in facing the challenges of the future.

The key driving forces for these changes are increasing demand for food, feed and other uses of biomass, such as energy, in the new bio economy, whereas land, water and other natural resources are limited (von Braun *et al.*, 2013; Mirzabaev *et al.*, 2014). Ongoing climate change is also likely to lead to higher frequency and magnitudes of extreme weather events such as droughts and floods, putting further negative pressure on crop and livestock productivity, especially in tropical and sub-tropical regions of the world, thereby presenting grave threats for global food security (Wheeler and von Braun, 2013; *von Braun, 2007*; Pingali, 2012). Furthermore, the rates of environmental degradation have already increased. For example, it is estimated that about a third of the global land area has experienced land degradation over the last three decades, affecting the livelihoods of more than three billion people (Le *et al.*, 2014), mostly the poor in developing countries (Nachtergaele *et al.*, 2010). Food, energy, land, water, mineral and financial markets have become increasingly interconnected. At the same time, the scientific and technological advances are facilitating groundbreaking changes in how economies would be shaped in a post-fossil fuel age, creating the basis for sustainable and green growth, for instance with alternative sources of energy, changing concepts of mobility and different lifestyles. A key feature of the bio economic system is that it values natural capital, including soil and water resources, as an essential building block of the economy, setting its management on the same level as the management of physical, human and other forms of capital (von Braun *et al.*, 2013).

These changes are bringing closer and integrating various facets of sustainable development – with its interlinked economic, social and environmental components – and magnifying both positive and negative externalities, making it quite impossible to solve the problems of sustainability in a piecemeal and isolated manner due to complex interactions and mutual dependencies. For example, the energy sector is becoming more water-intensive as bioenergy and hydropower diversify energy mixes. On the other hand, energy is essential to use water (lifting, pumping, desalination etc.), and food production is increasingly both water- and energy-intensive (Mirzabaev *et al.*, 2014). On the trade-off side, crop-based bioenergy and food production compete for land (Rathman *et al.*, 2010; Pilgrim and Harvey, 2010). The trade-off may cause conflicts, particularly when large-scale bioenergy competes with local water demand for food-crop production in water-scarce areas (Berndes, 2002). Demand for energy may contribute to deforestation, leading to soil erosion (Bazilian *et al.*, 2011), reducing crop productivity and, somewhat ironically, also reducing hydro-energy production through increased silting of dams (Nkonya *et al.*, 2014). In contrast, there are also numerous possibilities for synergies. For example, water operators spend about 70 percent of their revenue on energy costs. Mini-hydropower stations have been shown to reduce these energy costs for pumping stations by almost 80 percent (Kitio, 2013); the provision of cheaper micro-scale hydropower can help in the adoption of modern bioenergy technologies (Heltberg, 2004), and could potentially lower the demand for less sustainable traditional bioenergy use (cf. Mirzabaev *et al.*, 2014 for a review). Furthermore, the Nexus approach is critical to capture the inter-linkages between poverty, exclusion and ecology (von Braun and Gatzweiler, 2014). Global and local forces are necessitating the adoption of the Nexus approach for achieving sustainable development and poverty reduction in this ever more interconnected era.

Theoretical and conceptual underpinnings

Before discussing the theoretical and conceptual underpinnings of the Nexus, it should be stressed that the Nexus perspective as such is not a theory but an approach that provides a framework. Frameworks organize, form boundaries around the enquiry, set up general relationships among categories or dimensions, and define the scope and levels of the enquiry. They do not explain or predict; rather, they organize the diagnostic enquiry (Ostrom *et al.*, 1994). Theories and models are understood as "nested set[s] of theoretical concepts, which range from the most general to the most detailed types of assumptions made by the analyst" (Ostrom, 2005: 27). Theories often aim to explain particular parts of a framework, and therefore need to make assumptions about the patterns of relationships within frameworks.

The challenge to optimize public policy and public investments across sectors lies at the heart of the Nexus approach. A major objective of such public policies is to identify incentives and regulatory regimes for providing the optimal economic context for Nexus actions by the private sector, civil society and consumers, by removing market failures and minimizing transaction costs and negative externalities across the Nexus sectors. Transaction costs are the costs associated with participating in the market (Coase, 1937, 1960), i.e. executing a certain market transaction, and they relate to overcoming market imperfections. Transaction costs are usually categorized into three major types: 1) search and information costs; 2) bargaining costs; and 3) monitoring and enforcement costs (Furubotn and Richter, 2005; Koning, 2007). For example, a farmer wishing to sell her agricultural output would need to find customers and also know about the prevailing market prices, negotiate the transaction and make sure that the buyers pay for the sold output. The importance of transaction costs to Nexus thinking is straightforward. In contrast to the standard neoclassical approaches which might focus on profit maximization or cost minimization in specific sectors, new institutional economics, incorporating transaction costs in its analytical framework, is concerned more with the interactions between these Nexus sectors. This entails seeking to improve the overall Nexus functioning, i.e. the efficient and effective delivery of public and private goods and services, by minimizing transaction costs rather than just by optimizing separate individual sector functioning. In doing so, new institutional economics allows for bounded rationality of the actors and often-problematic property rights and contract enforcement (Williamson, 1996), also accounting for issues of political economy in governing the Nexus, which must take political transaction costs into consideration (e.g. negotiation cost, cost of coordination, time cost).

Another key concept in the Nexus framework is externalities. These are costs or benefits affecting an unrelated third party, who did not pay or get paid for these costs or benefits. Externalities could be negative or positive. A typical example of a negative externality is soil erosion. Unsustainable land management could lead to soil erosion, whereby soils are washed into rivers and streams and may lead to the siltation of downstream water reservoirs and eventually to decreased production of hydropower (Nkonya *et al.*, 2014). The land users who are responsible for this soil erosion usually do not compensate the hydropower generators for lost income, thereby imposing a negative externality on them. An example of a positive externality could be child vaccination. When vaccinated against infectious diseases, the benefit clearly accrues to the child herself and her parents, but also to other children and adults who are now less likely to get infected through this vaccinated child. In the absence of transaction costs, assignment of property rights could lead to internalization of externalities (Coase, 1960). However, transaction costs are ubiquitous, and thus there is a need for actions to create flexible institutions that minimize these transaction costs, and thereby also minimize negative externalities – for example, through collective action (Ostrom, 1990). The Nexus framework is such a governance mechanism that seeks to minimize transaction costs and negative externalities across the Nexus sectors. Another important premise of Nexus thinking is that such a framework would also allow for promoting synergies, i.e. boosting positive externalities across the Nexus sectors.

The key feature of our interconnected world with little or no remaining slack in the available natural resources is intensifying competition for these resources between different sectors, creating complex externalities. However, a major political economic problem is that Nexus-based governance structures operating across the sectors are lacking, and individual sectors remain entrenched within their boundaries, largely entrapped in the neoclassical economics-style drive to maximize their own individual benefits, often even to the detriment of aggregate social well-being. Such an "institutional tragedy" is perhaps the most important public policy challenge of the coming decades, and overcoming it is a major untapped benefit of the pursuit of a Nexus approach. The establishment of new Nexus-based institutions takes time, and in addition the degree of focus on Nexus intensity must follow optimality criteria. Nexus intensity must not be overdone. What is needed is not new and more discrete compartmentalization, i.e. better "silos", but flexible and resilient capacities, mechanisms and platforms allowing for a variety and diversity of interlinks depending on the nature of the targeted problems. Key opportunities that can now enable this more easily are the rapid deployment of information and communication technologies (ICTs), the increased transparency of Nexus governance domains through sharing and access to (real-time, big) data, much faster computing capabilities, e-government, creating conditions for institutional innovations, and cooperation.

The Nexus framework is called upon not only to bridge sectors that co-provide services to society better in conjunction than in isolation, but also to bridge scales. There may be Nexus interactions from global, national, regional, river basin and watershed to household levels, and even intra-household implications due to gender-differentiated impacts. For example, a shift from traditional fuel wood as a source of cooking and heating energy to clean, affordable and reliable renewable energy sources could particularly benefit and empower women within households, reducing their exposure to indoor air pollution and saving on their time otherwise spent on collecting fuel wood. Ideally, cross-scale analyses are needed to identify various Nexus trade-offs and synergies and undertake actions to minimize transaction costs across the Nexus sectors.

The conceptual diagram for the Water–Energy–Food Security Nexus is given in Figure 3.2. The key element of the Nexus is its focus on solving sustainable development problems with people at the centre of its preoccupations.

This people-centred approach can serve as a powerful rallying point for sustainability action, bringing social, economic and ecological issues jointly into focus. The major purpose is not to maximize or minimize certain individual element in the Nexus (for example, increasing "crop per drop" irrespective of anything else), but to optimize the whole Water–Energy–Food Security Nexus, with improvements in human welfare as the key yardstick measuring the success or failure of actions.

The Nexus approach includes working towards good governance, reducing corruption and providing legislative frameworks facilitating the mainstreaming of Nexus thinking into governance institutions and practice, with a critical role to be played by local communities and their involvement in managing Nexus sectors. The Food–Energy–Water Nexus is mostly local, though some elements are global, and hence appropriate actions are also needed at national, transnational and global levels. Such actions, at both local and higher levels, could be achieved through incentives, policy coherence mechanisms, institutional frameworks, education, information, research and empowerment. As more and more nations are decentralizing government, institutional opportunities for Nexus action in the delivery of services find a home at local level to manage water, energy and food security more effectively (Birner and von Braun, 2009).

Modelling the Nexus interactions

Potential Nexus benefits remain hidden, as long as a new generation of quantitative models is not created that explicitly incorporate Nexus features. This is a key challenge for Nexus science. Although there have recently been considerable developments in the conceptual formulation of Nexus thinking, the empirical applications through interconnected modelling of Nexus sectors are virtually non-existent. However, such modelling approaches are needed to provide an evidence-based assessment on the extent of actual trade-offs and synergies, and for simulating the systemic impacts of policy actions across Nexus sectors at global and national levels and across the sectors of economies. There exist various modelling tools working at various scales, such as integrated assessment models, partial and computable general equilibrium models, river basin level hydrological-economic models, household level bio-economic models, and agent-based models seeking to analyse behaviour at an (intra-) household level. However, in most cases these models are built on the

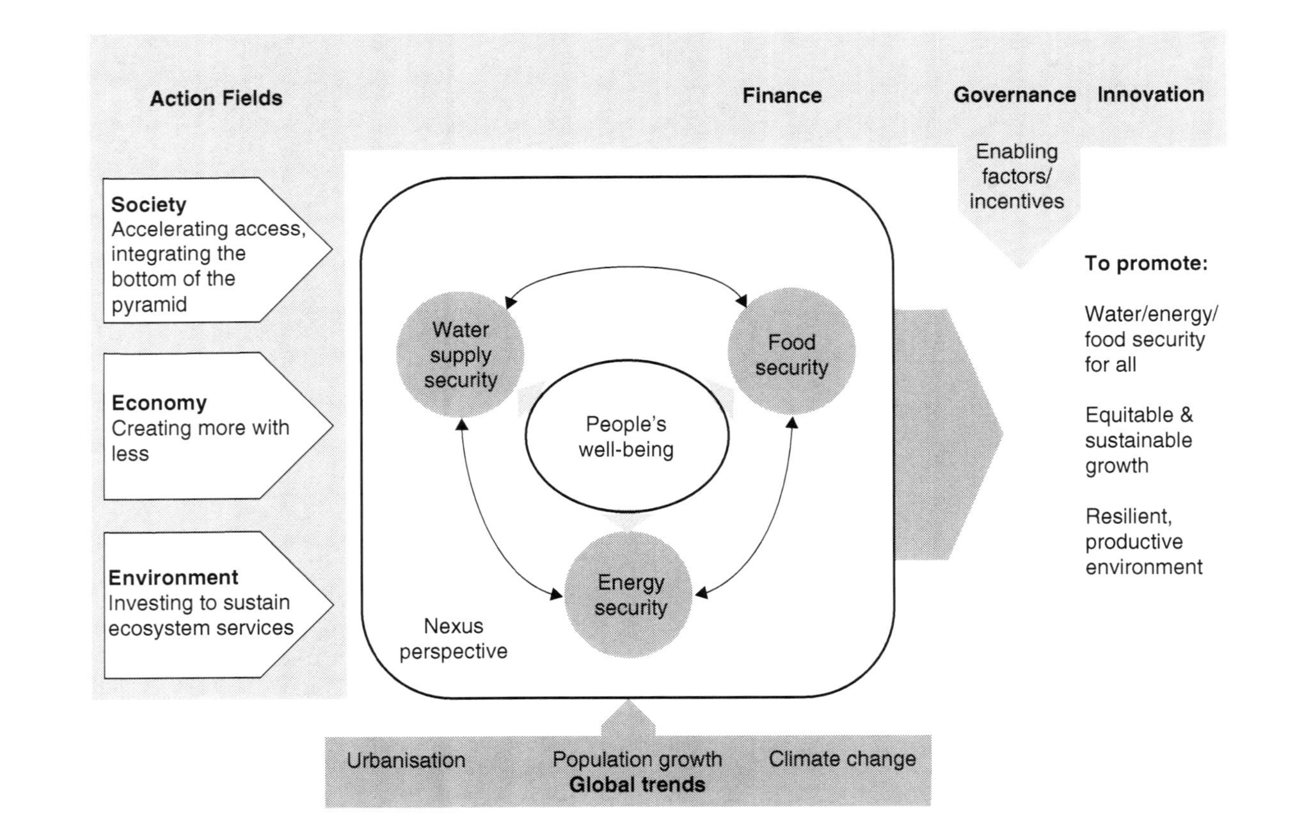

Figure 3.2 The Water–Energy–Food Security Nexus. Source: adapted from Bonn 2011 Conference on The Water, Energy and Food Security Nexus: Solutions for the Green Economy, Hoff (2011).

assumptions of profit maximization or cost minimization, say, in specific sectors of the Nexus, neglecting positive and negative externalities. As discussed above, Nexus thinking is based on new institutional economics, which, by incorporating transaction costs in its analytical framework, is seeking to improve the overall functioning of the set of sectors in the Nexus by minimizing transaction costs, minimizing negative externalities, and maximizing positive externalities. Thus the application of Nexus thinking would require re-equipping existing sets of models with Nexus features of transaction costs and externalities.

However, many such models running at the global level follow a system dynamics simulation approach, at very high levels of aggregation and often without covering all the Nexus sectors in sufficient detail. Integrated assessment models used to simulate future climate change scenarios could serve as typical examples. On the other hand, there exist several models covering specific interactions within selected Nexus sectors in more depth. For example, IFPRI's IMPACT model covers global food and water projections (Rosegrant, 2013).

Models can be classified into partial equilibrium and computable general equilibrium (CGE) models. The partial equilibrium models study specific economic sectors. For example, there have been studies conducted on energy and food production linkages based on partial equilibrium frameworks (Schneider *et al.*, 2007; Searchinger *et al.*, 2008; Chen *et al.*, 2011; Steinbuks and Hertel, 2012; Bryngelsson and Lindgren, 2013). Bryngelsson and Lindgren (2013), based on a partial equilibrium model, indicate that a large-scale introduction of bioenergy would raise food prices. Measures to limit the production of bioenergy crops on marginal lands can only partially reduce such food price increases at best, as there would be strong incentives to grow bioenergy crops on more fertile lands, ultimately leading to accelerated deforestation (ibid.).

On the other hand, computable general equilibrium (CGE) models consider the entire economy, thus capturing the interactions between sectors more accurately. To illustrate, the CGE models better reflect the effect of land reallocation from food production towards bioenergy through substitution effects, which further affect the relative prices of goods (food, energy, feed and fibre etc.). Accordingly, several studies have used CGE models to capture land use change (for example, Banse *et al.*, 2008; Hertel *et al.*, 2010; Bouët *et al.*, 2010). Despite these advantages of CGE models, they also have potential flaws. For example, the CGE model is not estimated but calibrated using a social accounting matrix (SAM) provided by the Global Trade Analysis Project (GTAP), with high levels of sectoral aggregation. Moreover, CGE models do not differentiate between direct and indirect land use changes.

If partial and CGE models are used to model the interactions within a sector or across all the sectors of the economy, some of the Nexus questions require spatially explicit representations in the models for more accurate insights. In particular, different river basins and watersheds within the same country may have varying resource scarcities, and thus different Nexus interactions. In many cases, one single river basin could span several countries. Such situations require the application of river basin models, which allow for more accurate inclusion of hydrological potentials and constraints into the models. For example, Ringler (2001), through development of an integrated economic-hydrologic river basin model for the Mekong River Basin in Southeast Asia, identifies there are significant trade-offs between off-stream and in-stream uses of water, and there is a need for appropriate institutions for enhanced economic efficiency in water allocation. However, the main challenge remains that most river basin models seek to maximize economic profits from water use in Nexus sectors where the water needs to be allocated to the sector with highest economic returns per unit of water (for example, Cai, 2008; Bekchanov, 2014), whereas the Nexus approach works on a conceptually more nuanced basis, seeking to minimize the transaction costs across all the Nexus sectors in an effort to raise the overall governance efficiency.

Nexus trade-offs and synergies relate not only to global or sectoral scales; they are also present at the household level. Moreover, this is the most crucial level for tracing the impacts of Nexus synergies and trade-offs on people's well-being. Agricultural household models have for a long time become workhorses for microeconomic analyses of production and consumption decisions, and for corresponding resource use and allocation by households (Singh *et al.*, 1986). Bio-economic modelling approaches, involving the coupling of crop modelling, household production and consumption decisions involving water, land and energy, have been extensively applied (Woelcke, 2003; Holden *et al.*, 2005; Djanibekov *et al.*, 2012), but here also the application of the Nexus framework would require a shift in its conceptual basis to embrace the principles of new institutional economics, including far-reaching impacts on the objective functions of the models and their assumptions about human behaviour.

The practical way forward is probably not to aim for all-comprehensive Nexus-mega models, but to enrich existing models step by step with Nexus elements, making them fit for purpose and thereby gaining additional insights.

WEF Nexus in science and practice

Over the last several centuries there have been ever-growing divisions in science, separating it into more and more distinct disciplines. Over time, with increased specialization and the development of distinct tools and methods, collaboration between different areas has become increasingly challenging. On the other hand, the problems of sustainable development have become increasingly interconnected, due to globalization, advancements in information and communication technologies, the increasing human footprint on the environment (the Anthropocene), and so on. Hence, to be able to address these interconnected problems there is a need to ensure collaboration among diverse disciplines and stakeholders, a requirement which is increasingly recognized in science communities in general, and in sustainability science in particular. WEF Nexus is an important example of such collaboration. Nexus science approaches are not a temporary whim in research re-designs, but an urgent necessity that is likely to become a growing and permanent feature of future science in general. Earlier scientific divisions into disciplines have also been replicated in the education system and eventually had an influence on the process of formation of public institutions. Nexus thinking in science should also be replicated through higher collaboration between existing public institutions and governmental departments. Political economy plays a key role in the success of the Nexus implementation. However, the political frameworks often do not always provide opportunities for maximizing Nexus synergies.

The success of the WEF Nexus depends on the success of actions in Nexus governance, trans-sectoral actions, and Nexus science.

- *Nexus governance* requires well-functioning decentralization. Fortunately, there is a positive worldwide trend toward improved decentralization. Three elements of effective decentralization matter for Nexus governance are: 1) administrative decentralization; 2) political decentralization; and 3) fiscal decentralization (Birner and von Braun, 2009). For decentralization to work successfully, usually all three elements need to be present (ibid.). Key aspects for the success of Nexus-inspired approaches to public policy are their ability to promote institutional innovations and cooperation, as well as technological innovations across the Nexus sectors. In this regard, the emergence of modern bioeconomy as a core element of "green growth", urges for accelerated implementation of Nexus-inspired policy actions.[2] Technical innovation is an ingredient of sustainable bio economy and bioenergy development, as it may help to minimize the risks that may arise from trade-offs between bioenergy and food security through increasing efficiency and efficacy of resource use.
- *Trans-sectoral actions* include those areas with multi-stakeholder systems, for example, bio economy (connecting water, food and energy in "value webs").
- *Nexus-oriented science* in such areas as comprehensive footprint analyses (e.g. global externalities of scarce water footprints in multi-country models) is critical. Similarly, comprehensive food security and climate change research, with a shift from the current 85 percent focus on production effects and neglect of access, volatility, extreme events and nutrition (see Wheeler and von Braun 2013), should be high on the science agenda, as well as information and behavioural change support in complex Nexus environments, e.g. water–sanitation–agriculture linkages in multi-use water systems, and the integration of Nexus with resilience approaches.

Conclusions

The Water–Energy–Food (WEF) Security Nexus is a conceptual framework that recognizes intricate interconnections among these broad sectors and seeks to develop joint solutions that mitigate the trade-offs and promote synergies among them. The WEF Nexus is not only about sectoral practices, but also about how scientific research is conducted within these Nexus areas. Nexus thinking is conceptually based on new institutional economics. Thus, in contrast to standard neoclassical approaches, which focus on profit maximization or cost minimization, the new institutional economics is concerned more with the interactions between these Nexus sectors. Improving the overall Nexus governance structure by minimizing transaction costs, rather than optimizing separate individual sector functioning, is at the core of Nexus institutional research. There may be Nexus interactions at different scales, from global to household levels. Ideally, cross-scale analyses are needed to identify various Nexus trade-offs and synergies and undertake actions to minimize the transaction costs across Nexus sectors. Although there have recently been advances in the conceptual formulation of Nexus thinking, empirical applications derived from the interconnected modelling of Nexus sectors are virtually non-existent. The application of Nexus thinking in quantitative modelling would require substantial innovations. Moreover, Nexus science requires a rethinking of science organization and management, including also science training, scientific publishing and science funding. Nexus practice would require optimizing investments in Nexus rather than maximizing them, an optimal balance between Nexus and sector actions, assessing the transactions costs associated with Nexus governance, and realism regarding Nexus political economy.

Notes

1 The lack of a wide occurrence in the literature should not be confounded with absence of such practices in earlier periods. Arguably, interdisciplinary thinking was already present in some form in ancient Greek philosophy and medieval scientific practices, especially since many of the disciplines which are distinct now were then considered to be unitary – for example, economics was a part of philosophy. What is different now is the incomparably more complex depth and the often clear distinctiveness in tools and methods from each individual discipline, making interdisciplinary research at the same time more challenging, in terms of integrating diverse disciplinary agendas and approaches, and more rewarding, in terms of accurate and comprehensive research insights.

2 The aggregate of all industrial and economic sectors and their associated services, which produce, process or in any way use biological resources (BioÖkonomieRat, 2009).

References

Banse, M., van Meijl, H., Tabeau, A. and Woltjer, G. (2008) 'Will EU biofuel policies affect global agricultural markets?' *European Review of Agricultural Economics*, 35(2), 117.

Bazilian, M., Rogner, H., Howells, M., Hermann, S., Arent, D., Gielen, D., Steduto, P. *et al.* (2011) 'Considering the Energy, Water and Food Nexus: Towards an Integrated Modelling Approach', *Energy Policy*, 39(12), 7896–7906. doi: 10.1016/j.enpol.2011.09.039

Bekchanov, M. (2014) *Efficient Water Allocation and Water Conservation Policy Modeling in The Aral Sea Basin*. PhD diss., Universitäts-und Landesbibliothek Bonn, Germany.

Berndes, G. (2002) 'Bioenergy and water – the implications of large-scale bioenergy production for water use and supply', *Global Environmental Change*, 12. Available online at: https://www.researchgate.net/publication/222709547_Bioenergy_and_Water_The_Implications_of_Large-Scale_Bioenergy_Production_for_Water_Use_and_Supply

BioÖkonomieRat (2009) *Combine disciplines, improve parameters, seek out international partnerships. First recommendations for research into the bio-economy in Germany BÖR, Berlin.* Available online at: http://biooekonomierat.de /fileadmin/Publikationen/Englisch/BOER_Recommendation02_research.pdf

Birner, R. and von Braun, J. (2009) 'Decentralization and public service provision: A framework for pro-poor institutional design', in E. Ahmad and G. Brosio (Eds) *Does Decentralization Enhance Service Delivery and Poverty Reduction?* Cheltenham: Edward Elgar, 287–315

Bouët, A., Dimaranan, B.V. and Valin, H. (2010) *Modeling the global trade and environmental impacts of biofuel policies.* IFPRI Discussion Paper (01018), International Food Policy Research Institute.

Bryngelsson, D.K. and Lindgren, K. (2013) 'Why large-scale bioenergy production on marginal land is unfeasible: A conceptual partial equilibrium analysis', *Energy Policy*, 55, 454–466.

Cai, X. (2008) 'Implementation of holistic water resources-economic optimization models for river basin management: Reflective experiences', *Environmental Modeling and Software*, 23(1), 2–18.

Chen, X., Huang, H., Khanna, M. and Önal, H. (2011) *Meeting the Mandate for Biofuels*. NBER Working Paper 16697, 1–60.

Coase, R.H. (1937) 'The nature of the firm', *Economica*, 4(16), 386–405.

Coase, R.H. (1960) 'The problem of social cost', *Journal of Law and Economics*, 3, 1.

Djanibekov, U., Khamzina, A., Djanibekov, N. and Lamers, J. (2012) 'How attractive are short-term CDM forestations in arid regions? The case of irrigated croplands in Uzbekistan', *Forest Policy and Economics*, 21, 108–117.

Furubotn, E.G. and Richter, R. (2005) *Institutions and economic theory: The contribution of the new institutional economics*. Ann Arbor: University of Michigan Press.

Heltberg, R. (2004) 'Fuel switching: Evidence from eight developing countries', *Energy Economics*, 26(5), 869–887.

Hertel, T., Golub, A., Jones, A., O'Hare, M., Plevin, R. and Kammen, D. (2010) 'Effects of US maize ethanol on global land use and greenhouse gas emissions: Estimating market-mediated responses', *BioScience*, 60, 223–231.

Hoff, H. (2011) *Understanding the Nexus*. Background Paper for the Bonn 2011 Conference: The Water, Energy and Food Security Nexus. Stockholm Environment Institute, Stockholm.

Holden, S.T., Bekele, S., Shiferaw, B.A. and Pender, J. (2005) *Policy analysis for sustainable land management and food security in Ethiopia: A bioeconomic model with market imperfections.* Washington, DC: International Food Policy Research Institute.

Kitio, V. (2013). *Energy and Water Nexus Opportunities and technical solutions to address poor access to water and energy through better understanding of the water and energy nexus.* Global

Conference: "Rural Energy Access: A Nexus Approach to Sustainable Development and Poverty Eradication". Addis Ababa, Ethiopia, 4–6 December 2013.

Koning, J. (Ed.) (2007) *The evaluation of active labour market policies: Measures, public private partnerships and benchmarking*. Cheltenham: Edward Elgar Publishing.

Le, Q.B., Nkonya, E. and Mirzabaev, A. (2014) *Biomass productivity-based mapping of global land degradation hotspots*. ZEF-Discussion Papers on Development Policy (193).

Miller, R.C. (1982) 'Varieties of interdisciplinary approaches in the social sciences: A 1981 overview', *Issues in Integrative Studies*, 1, 1–37.

Mirzabaev, A., Guta, D.D., Goedecke, J., Gaur, V., Börner, J., Virchow, D., Denich, M. and von Braun, J. (2014) *Bioenergy, food security and poverty reduction: Mitigating tradeoffs and promoting synergies along the Water-Energy-Food Security Nexus*. ZEF Working Paper Series (No. 135). Bonn, Germany.

Mollinga, P.P. (2010) *Boundary concepts for interdisciplinary analysis of irrigation water management in South Asia*. ZEF Working Paper Series (No. 64). Bonn, Germany.

Nachtergaele, F., Petri, M., Biancalani, R., van Lynden, G. and van Velthuizen, H. (2010) *Global Land Degradation Information System (GLADIS), Beta version: An information database for land degradation assessment at global level*. Land Degradation Assessment in Drylands Technical Report, no. 17. FAO, Rome, Italy

Nkonya, E., Gerber, N., Baumgartner, P., von Braun, J., de Pinto, A., Graw, V., Kato, E., Kloos, J. and Walter, T. (2014) *The economics of desertification, land degradation, and drought – Toward an integrated global assessment*. ZEF-Discussion Papers on Development Policy No. 150, Center for Development Research (ZEF), Bonn, Germany

Ostrom, E. (1990) *Governing the commons: The evolution of institutions for collective action*. Cambridge: Cambridge University Press.

Ostrom, E. (2005) *Understanding institutional diversity*. Princeton/Oxford: Princeton University Press.

Ostrom, E., Gardner, R, and Walker, J. (1994) *Rules, games, and common pool resources*. Ann Arbor: University of Michigan Press.

Pilgrim, S. and Harvey, M. (2010) 'Battles over biofuels in Europe: NGOs and the politics of markets', *Sociological Research Online*, 15. Available online at: http://www.socresonline.org.uk/15/3/4.html

Pingali, P. (2012) 'Green Revolution: Impacts, limits, and the path ahead', *Proc Natl Acad Sci USA*, 109, 12302–12308.

Rathman, R., Szklo, A. and Schaeffer, R. (2010) 'Land use competition for production of food and liquid biofuels: An analysis of the arguments in the current debate', *Renewable Energy Online*, 35(1), 14–22.

Ringler, C. (2001) *Optimal allocation and use of water resources in the Mekong River Basin: Multi-country and intersectoral analyses*. Development Economics and Policy Series Vol. 20 (PhD dissertation, Bonn University). Frankfurt: Peter Lang Verlag.

Ringler, C., Bhaduri, A. and Lawford, R. (2013) 'The nexus across water, energy, land and food (WELF): Potential for improved resource use efficiency?' *Current Opinion in Environmental Sustainability*, 5(6), 617–624.

Rosegrant, M.W. (2013) *International model for policy analysis of agricultural commodities and trade (IMPACT): Model description*. IMPACT Development Team.

Schneider, U.A., McCarl, B.A. and Schmid, E. (2007) 'Agricultural sector analysis on greenhouse gas mitigation in US agriculture and forestry', *Agricultural Systems*, 94(2), 128–140.

Searchinger, T., Heimlich, R., Houghton, A., Dong, F., Elobeid, A., Fabiosa, J., Tokgoz, S., Hayes, D. and Yu, H. (2008) 'Use of US croplands for biofuels increases greenhouse gases through emissions from land-use change', *Science*, 319, 1238–1240.

Singh, I., Squire, L. and Strauss, J. (1986) *Agricultural household models: Extensions, applications, and policy*. Baltimore: Johns Hopkins University Press.

Steinbuks, J. and Hertel, T. (2012) *The optimal allocation of global land use in the Food-Energy-Environment trilemma*. Working Paper No. 12-01.

von Bertalanffy, L. (1956) 'General system theory', *General Systems*, 1(1), 11–17.

von Braun, J. (2007) *The world food situation: New driving forces and required actions*. Washington, USA: IFPRI.

von Braun, J. and Gatzweiler, F. (Eds) (2014) *Marginality – Addressing the Nexus of Poverty, Exclusion and Ecology*. Netherlands: Springer.

von Braun, J., Gerber, N., Mirzabaev, A. and Nkonya, E. (2013) *The economics of land degradation*. ZEF Working Papers, No. 109. Bonn, Germany.

Wheeler, T. and von Braun, J. (2013) 'Climate change impacts on global food security', *Science*, 341(6145), 508–513.

Williamson, O.E. (1996) *The mechanisms of governance*. Oxford: Oxford University Press.

Woelcke, J. (2003) *Bio-economics of sustainable land management in Uganda*. Oxford: Peter Lang.

4 Global risks and opportunities in food, energy, environment and water to 2050

R. Quentin Grafton

Introduction[1]

In this chapter I reflect on where the world was in 1968, where it is heading in the context of the Food, Energy, Water and Environment Nexus, and how researchers should respond. Much has changed since 1968 when the Apollo 8 mission was the first to journey around the moon. Not least, the world's population was then 3.6 billion. It is now more than twice as large, and is projected to be 9.6 billion by 2050.

The year 1968

1968 was arguably the year that hope died for many who wanted a better world. It could be described as the year of the 'perfect storm'. Robert (aka Bobby) Kennedy, Democratic Presidential candidate and brother of President John F. Kennedy, and also the Reverend Martin Luther King Junior were both assassinated in that critical year.

These two assassinations, the ending of the Prague Spring by Soviet tanks, the Paris student riots, a worsening conflict in Nigeria with the Biafran War and a low point in the Chinese Cultural Revolution all happened in 1968. Such events changed the *Zeitgeist* of the time. This, I contend, is why the picture of a small Earth alone in space captured by the Apollo 8 mission – the notion of a 'spaceship earth' – captured the imagination of all who saw it then, and all who see it today. It unites us all.

It is, therefore, fitting to recall Bobby Kennedy, who said:

> There are those who look at things the way they are, and ask why... I dream of things that never were, and ask why not?

I am inspired by this message of hope. It is not a question of whether the glass is half empty or half full, but more a question of what we are going to do about it. Instead of simply looking at the glass, we can play our part and actually fill it.

In his Nobel acceptance speech to the Swedish Academy the writer John Steinbeck, the author of "The Grapes of Wrath", stated:

Man himself has become the greatest challenge and our only hope.

We can no longer afford to be spectators to what is happening in the world today; we need to become actors that play an active role in what happens on the global stage.

To the present

Despite the chaos of 1968, in the following years the world successfully overcame a pending global food crisis despite the fact that the human population was then growing at close to its fastest rate ever. Even given this formidable challenge, global food supplies tripled and world per capita incomes more than doubled in real terms. Few people back in 1968 would have believed that such successes were possible, since the world then faced a Cold War that was at its height and the risks of nuclear conflict seemed so near.

The very positive changes witnessed over the past fifty years or so did not just happen by accident. They were caused by the deliberate actions of individuals in both the private and public sectors. Investments were made in high-yielding crop varieties, knowledge was created and diffused, and gradually people were given opportunities in terms of better education, health care and employment.

It is worth considering these human triumphs when thinking about what we must do in the next thirty-five years as we face the human-created challenge of climate change, the need to produce at least 60 percent more food (FAO, 2014) and how we can respond to a projected increased water demand of at least 55 percent (WWAP, 2014) and a primary energy demand of 40 percent by 2050 (IEA, 2014), but without damaging the soils and resources we need to feed the world.

The perfect storm

I contend that our past successes made many of us complacent in the 1990s and the early 2000s. Seemingly, somehow it was assumed that business as usual would continue to deliver the food and environmental services that we all need, now and forever.

The rude awakening came exactly forty years after 1968. In 2008–09 the world experienced its worst economic recession in seventy years with the Global Financial Crisis (aka the Great Recession). It showed that very large and unexpected shocks can happen almost ‘out of the blue’ and be enormously damaging. Equally as important, it showed that driving our planet on ‘cruise control’ is simply not good enough, and that past ways of decision making and understanding of risks had spectacularly failed.

Immediately before the global economic crisis the world had also experienced a food crisis. Average food prices rose by around 80 percent on average from 2005 to 2008, primarily as a result of high energy prices but also in part due to misguided policies to subsidize biofuel production. Indeed, in the absence of biofuel production the global food price in 2008 would have been 23 percent lower than it was actually (To and Grafton, 2015). The end result of both higher energy prices and biofuel subsidies was that globally an additional 100 million people became chronically undernourished.

These price increases, and another price spike in wheat in 2010–11, were contributory factors to the Arab Spring that got under way in 2011. Just as in the French Revolution 220 years previously, very high food prices generated their own political and socio-economic risks.

The food price spikes of 2008–2011 show the systemic risks between food and energy. Recent droughts, such as Australia's Millennium Drought that ended in 2010 or the California Drought which has been four years developing and is still ongoing in 2015, illustrate our critical dependence on water for food production and environmental services.

1968 was the year that the phrase 'Green Revolution' was popularized in reference to the way that higher-yielding varieties, coupled with greater inputs, could transform food production. One of the fathers of the Green Revolution, Norman Borlaug, who did pioneering work in plant breeding to produce high-yielding wheat varieties, was awarded the Nobel Prize for his contributions to world peace as a result of his efforts to increase global food supplies.

Unlike 1968, in 2015 we are approaching, or perhaps have surpassed, key planetary boundaries or limits. Simply put, it is not physically possible to increase food supplies at the expense of extracting water at an increasing rate, or by causing irreparable soil degradation in many of the key food growing regions of the world (Beddington, 2009). Instead, we need an alternative vision that has been called 'sustainable intensification' (Godfray, 2015).

I interpret sustainable intensification as increasing food production, but without using more arable land or inputs (such as water) that would irreparably damage critical environmental systems or degrade our long-term ability to produce food.

Full consideration of the systemic risks across food, energy and water and their collective impacts on the environment is achievable. It is no pipe dream, but it will not happen simply by wishing it, by moral exhortation to become vegans, or by resting on past laurels.

Too often we wring our hands, take amazing pictures to illustrate our despair, make documentaries, swarm together on social media and care loudly and publicly. We have telegraphed the problem to the world.

Navigating the perfect storm

The global food, energy and water risks of today demand nothing less than a 'paradigm shift' in how we see the world, and how we make decisions. 'Silo thinking', which may have delivered in the past, is simply not good enough in a highly connected world where a drought in China can trigger a revolution in Egypt.

There is a better way. It is not 'tears for fears', and neither is it more hand-wringing or even food aid, however well meaning. What is required are ways to help decision makers make better decisions, whether farmers, catchment managers or presidents.

My colleagues and I in the Food, Energy, Environment and Water (FE2W) Network have developed a Risk and Opportunities Assessment for Decision-making (ROAD) which can be applied at different scales and scopes, and is intended to improve decision making by accounting for systemic risks. It is not an optimization framework, but instead provides ways of assessing risks that explicitly incorporate the decisions that need to be made by individuals, businesses and governments.

Our approach includes:

1 Delineation of the decision space and questions;
2 Evaluation of threats and their triggers;
3 Causal risk and opportunities assessment;
4 Analysis of decision options for controls and ways to mitigate consequences, including a summary and justification of the selected decisions; and
5 Update and review following the decisions and their implementation.

The time and effort needed to apply our framework will depend on the temporal and geographical scale of the decisions at hand. For a subsistence farmer it may be a simple and virtually costless process, but one that would provide a more systematic way to assess and mitigate risks. By contrast, for a state water department responding to food and water risks along the Ganges, the process would be much more involved, considering multiple triggers and threats and risks and delivering a range of decision controls and options.

Whatever the scale or challenge, the value added by the Risk and Opportunities Assessment for Decision-making (ROAD) lies in providing a way for decision makers to make causal, risk-based decisions. This is urgently needed in the food, energy, environment and water space.

The FE2W Network plans to work at different scales in key parts of the world: the Volta Basin and the Nile Basin, both in Africa; the Ganges–Brahmaputra–Meghna in South Asia; the Mekong in South-East Asia; the Colorado in the US and Mexico; and the Murray–Darling in Australia.

Our efforts to navigate the perfect storm need the support of donors, thinkers and doers. Collectively we need to, and we must, build durable partnerships and the means to scale up the ROAD framework to generate improved outcomes on the ground and around the world.

This is a collective effort. It requires that we manage and mitigate the systemic risks across food, energy, the environment and water so that we can promote a sustainable and prosperous future for all of humanity.

The way forward

There is no question that the world faces major challenges over the coming decades in the context of food, energy, the environment and water. However, just as in 1968, when it all seemed 'too hard' and when many believed that the world was heading towards mass starvation, we collectively turned the ship around in very stormy waters.

The time for Brothers Grimm fables is over. The time for action is at hand. Many opportunities exist, and better decision making can deliver solutions to the challenges that we face. We know what to do and where to start to work.

Humanity proved the 'food doom mongers' wrong in 1968. This was done by using more land, a lot more inputs, especially fertilizers and water, improving technologies, and with a single-minded focus on production (Pingali, 2012). In the next thirty-five years we must act differently. Increasing food supplies at the expense of irreversible soil loss or the mining of key aquifers, or much greater greenhouse gas emissions, is no longer sustainable – if it ever was.

Now an alternative path is required. This pathway to the future demands better decision making that 'connects the dots' across food, energy, environment and water. It requires decisions that fully account for the value chains 'from the farm gate to the dinner plate'. It needs integrated assessment and management to promote resilient food–energy–water systems that can withstand unexpected and negative shocks, and decisions that do not compromise future sustainability.

Better risk-based decisions are needed if we are to 'navigate the perfect storm' that lies ahead of us to 2050 and beyond. This is not a journey we can take by ourselves; we must work together.

Conclusion

We can continue to feed and clothe the world, but not in ways that compromise our common future. Instead, we must act to ensure our food and natural systems are resilient to shocks, and that we are responsive to the opportunities of the unexpected.

What is required is better decision making that focuses on risks and opportunities and at multiple scales and contexts. This requires a paradigm shift in how decisions are made, what information is gathered and how it is used. It also demands a 'Nexus' approach to responding to systemic risks, encompassing multiple disciplines and meaningful partnerships between researchers, decision makers and those who implement the decisions.

Note

1 This chapter is based on a speech given by the author at the Policy Forum (www.policyforum.net) Lunch on 22 April, 2015 at the Australian National University, Canberra.

References

Beddington, J. (2009) *Professor Sir John Beddington's Speech at the SDUK09 Conference, London*. Available online at: www.govnet.co.uk/news/govnet/professor-sir-john-beddingtons-speech-at-sduk-09

FAO (Food and Agriculture Organization) (2014) *The Water–Energy–Food Nexus: A new approach in support of food security and sustainable agriculture*. Rome: FAO.

Godfray, C. (2015) The debate over sustainable intensification, *Food Security* (Impact Factor 1.5), 4, 7(2), 199–208. Available online at: www.researchgate.net/publication/275226297_The_debate_over_sustainable_intensification

IEA (International Energy Agency) (2014) *World Energy Outlook*. Paris: IEA.

Pingali, P. (2012) Green revolution: Impacts, limits, and the path ahead, *Proceedings of the National Academy of Sciences of the United States of America*, 109(31), 12302–12308.

To, H. and Grafton, R.Q. (2015) Oil prices, biofuels production and food security: Past trends and future challenges. *Food Security*, 7(2), 323–336. Available online at: www.researchgate.net/publication/282842824_Oil_prices_biofuels_production_and_food_security_past_trends_and_future_challenges

WWAP (World Water Assessment Programme) (2014) The United Nations World Water Development Report 2014: Water and Energy, 1. Paris: UNESCO.

5 The Water, Energy And Food Nexus and ecosystems

The political economy of food and non-food supply chains

Tony Allan and Nathanial Matthews

Introduction

The ways in which the Water, Energy and Food Nexus, hereafter the Nexus, is managed and water and energy are traded and consumed globally have important implications for ecosystem stewardship. Water, energy and food are primarily managed in dynamic multiscale supply chains. Supply chains comprise complex networks of organizations, individuals, activities, technologies and resources. They produce commodities and provide services to communities by sourcing inputs (Handfield and Nichols, 1999). Supply chains are a useful organizing construct that can help illuminate the political economic forces that influence how ecosystems are valued in the Nexus because they are governed by prices, market rules and public regulation that are both political and economic. Consumers, the private sector, governments and economies depend on effective supply chains for the delivery of cheap water, energy and food in order to function. They all have vested interests in influencing how these supply chains operate.

In meeting the needs of the political economies of human communities, supply chains place variable demands, with varying levels of intensity, on natural ecosystems. The players who know most about these market supply chains are those who operate them. However, as yet levels of ecosystem awareness are not high and ecosystems are not considered important among producers, traders, processors, marketing agents and consumers. In the past decade, major private sector players have recognized Nexus hotspots where water, energy or both have impaired their capacity to operate profitably in particular locations. These hotspots involve water, energy and food at a small scale and do not involve the Nexus that combines natural inputs across regional or global supply chains. At the same time, because supply chains operate in very complex and hidden ways the science community is finding it extremely difficult to establish the metrics that expose the impacts of our supply chains on natural ecosystems, and determine whether the supply chains are sustainable.

The importance of supply chains in managing the world's water, food and energy resources, and hence this important global Nexus, cannot be overestimated. Food supply chains, for example, account for 92 percent of the water consumed in the world's economies and about 20 percent of the energy (Hoogeveen, 2015). This means that the majority of the world's human water consumption is managed by millions of individual farmers. These farmers include subsistence farmers, commercial family farmers and corporate farmers. Many grow food for themselves in low input, low output systems. An increasing proportion sell food into local or global markets.[1] Scarcely any of the water managed and mismanaged by farmers is priced as an input. The energy inputs are for the most part only captured on commercial farms. At the same time, governments often subsidize the energy used to pump and deliver water for irrigation. This practice adds to the long list of factors that make the market prices of food and fibres very poor indicators of the ecosystem costs of production across the Nexus.

The non-food and fibre supply chains also require water and energy services in providing industrial, domestic and transport services. These supply chains account for about 8 percent of the water consumed globally, but for a very high proportion – about 80 percent – of the energy (Allan *et al.*, 2015). Water and energy consumption in non-food and fibre supply chains provide services that make the lives of consumers at home comfortable, their daily journeys possible and quick, and meet their myriad non-food and fibre consumption choices. These food and non-food supply chains and the markets in which the Nexus operates are shaped by very powerful invisible and unaccounted forces.

The world's food and energy markets and supply chains that interact across the Nexus are influenced by political economic mechanisms such as subsidies, lobbying by farm and energy industry interests, and state-led public policies. All these mechanisms are shaped by emotional imperatives from the domestic to the international levels (Goodman *et al.*, 1990; Berry, 1974). Employment, livelihoods and food poverty have always trumped ecosystem service priorities, even though they rely on them to function. These markets, their consumers (us) and the agri-food businesses and energy corporates that control the majority of the transactions within them have inadequate and dangerously blind accounting and reporting systems. Current reporting and accounting systems cannot effectively account for the impact of the consumption of water and energy in food and fibre production on the ecosystem services of water, the atmosphere and biodiversity. The absence of well-functioning accounting and reporting systems means that the farmers who produce food and fibre and the providers of energy services cannot be incentivized to steward nature's essential ecosystems by market mechanisms.

In this chapter, we argue that the Nexus concept must incorporate a strong ecosystem stewardship component. It must shift from a focus on addressing hotspots, which is the approach that currently dominates Nexus analysis, to one that examines the uses of water, energy and food production and consumption in supply chains across the Nexus, and their impact on ecosystems. To date the hotspots approach has stimulated the engagement of the private sector supply chain players in Nexus debates, since it is these professionals who have to confront the consequences of the scarcity of water and/or energy in their operations.

Throughout this chapter we use a political economy analysis to show how renewable and non-renewable water and energy resources are consumed and managed in long-established private sector supply chains in the food sector. These supply chains operate in, and are driven by, markets that are greatly distorted by allocative and management practices that do not account for the value of water or of the other ecosystems – the atmosphere and biodiversity – on which they depend. We then explore the consequences of these political economy distortions in Nexus contexts. We maintain that the upshot of Nexus systems that do not value ecosystems is that they tend to address collapses or weakness along global supply chains by looking at smaller scale hotspots, as opposed to examining these problems through a supply chain lens.

The focus on hotspots is driven by a deficient valuation of ecosystem services both locally and globally across the Nexus, and it contradicts one of the key features of the Nexus concept that promotes a more interdisciplinary and inclusive approach to water, energy and food security. We show that ecosystems are inadequately valued and misunderstood, before suggesting potential leverage points that could promote ecosystems stewardship, which, it is argued, is a fundamental element of the Nexus concept.

Brief literature review and methodology

The ideas of interdependencies, trade-offs and interdisciplinary approaches within the Nexus are not new. Indigenous people, farmers and modellers, to name but a few, have recognized their importance for a long time. However, the emergence of these terms grouped together under the concept of the Water, Energy and Food Nexus, hereafter the Nexus, is more recent. Scott *et al.* (2014) have traced the early roots of the Nexus concept in philosophy in the late 1920s through to its first links to natural resource management in the Food–Energy Nexus programme of the United Nations University in 1983, and then to its focus on water–energy aspects (cf. Durant and Holmes, 1985; Ingram *et al.*, 1984; Gleick, 1994) as a response to the global political and economic forces driving energy price spikes across the US.

However, the rise to prominence of the Nexus on the global stage arguably came with the global energy and food crisis of 2007–08. These food and energy price spikes were a poignant reminder of the fragility of the complex web of supply chains and ecosystem services. States, markets, farmers, energy producers and others have to operate in this web to feed hungry consumers and businesses around the globe. In recognition of the potential impact that political, economic and natural shocks such as floods and droughts could have on the global economy, in 2008 the World Economic Forum (WEF), which includes a wide range of business leaders and key players from across the Nexus, initiated a debate 'to raise awareness and develop a better understanding of how water is linked to economic growth across a nexus of issues' (WEF, 2011).

Water was used as the focal point in the WEF Nexus discussions because of its key role in both food and energy production. The WEF meeting was followed in 2011 by the Bonn conference, funded by the German government, that examined vulnerabilities and trade-offs from social, ecological and economic dimensions across the Nexus. The Nexus concept's wide adoption by NGOs, governments and academics, as evidenced by websites, reports and journal publications since 2011, demonstrates the influence this 'new' lexicon has had on sustainable development dialogues.

The political economy approach adopted in this chapter has already been deployed to analyse Nexus issues at varying scales. Allouche et al. (2015) argue that the Nexus does not adequately engage with the global political economy of energy and food. They argue that policy documents tend to promote economic and technological solutions that ignore inequalities across the global political economy. The Nexus concept recognizes that trade-offs may be used by decision-makers to justify negative environmental impacts as an unavoidable cost in order to provide a version of water, food and energy security. Dupar and Oates (2012) also use global political economic considerations to argue that the Nexus approach must take account of rights-based approaches and use transparent negotiations to ensure that resources are not commodified in ways that ignore the environmental costs of their consumption. Although we do not examine rights as part of our analysis, research is also needed to show how the food and non-food supply chains we analyse in this chapter impact labour and livelihoods.

Valuing ecosystems in the Nexus: Supply chains and the political economy

A major impediment to the operationalization of the Nexus approach is the non-valuation or the mis-valuation of water and energy. Water is usually priced in water and sewage supply chains as well as in energy supply chains. These supply chains account for about 8 percent of global water consumption. Water is always priced as an input in industrial uses including in the provision of energy services, and it is usually priced in the provision of domestic and urban water services. However, these three areas only account for approximately 8 percent of the water consumed around the globe.

The other 92 percent of water consumption, mainly consumed on farms in the production of food and fibre, is unpriced or inadequately priced. All green water – that is, the effective rainfall available for crops, accounting for about 70 percent of food and fibre production – is managed as if it were a free good.[2] The costs of delivering blue water – that is, surface and groundwater – associated with the other 30 percent of food and fibre production are not effectively captured. Nor are the negative impacts of mismanaging the use of blue water.

Nature's solar energy is also taken for granted and is not attributed a value, but the energy derived from renewable and non-renewable natural resources consumed in all economic sectors in both food and non-food supply chains is priced. Unfortunately these prices are seriously distorted by subsidies and taxes that make it extremely difficult to track the costs of energy inputs and the externalities, especially the environmental externalities, of their production and use. A recent study by the International Monetary Fund (IMF), for example, found that fossil fuel subsidies equate to $5.3tn a year, the equivalent of $10 million a minute every day or more than the total health spending of all the world's governments (Coady *et al.*, 2015; Carrington, 2015). Because of such market distortions, the accounting and reporting systems within them do not capture the cost or value of the environment or of the ecosystem services, which are vital to their operation.

Figure 5.1 provides an organizational architecture for understanding the Nexus. It demonstrates the importance of food supply chains, water and sewage supply chains and energy supply chains, labelled the Sub-Nexi, in delivering systems of provision to the world's political economies and the Nexus. As mentioned above, it is important to note that the sub-nexi, also understood as food and non-food supply chains, use water and energy differently. The majority of water is consumed in the food supply chain and the majority of energy is consumed in the energy supply chains. Although the WASH supply chain is important for life, it is not a significant global consumer of water or energy. Neither the sub-nexi nor the Nexus account for the ecosystem services that are essential for their operation. In an ideal world, the Nexus would provide a sound approach to allocating and managing water and energy in supply chains. Such an approach would recognize the importance of reciprocation and mutuality in natural resource consumption and ecosystem stewardship.

<table>
<tr><td colspan="3">The Nexus
Water/Energy/Food
Does not value natural ecosystems and is subservient to the political economy
Most Nexus analysis has attempted to conceptualize Nexus processes at this level, but it has proven difficult to understand and analyse the influence of the political economy and the impact on ecosystems at this scale.
In practice the Nexus is populated by three non-integrated market supply chains, shown below, that help to analyse the forces influencing the Nexus and its impacts on ecosystems. Those operating these separate supply chains have not yet recognized or adopted principles and practices of mutuality. They are unlikely to do this until market systems introduce accounting and reporting rules, recognizing the value of natural ecosystems that incentivize such behaviour.</td></tr>
<tr><td colspan="3">The Sub-Nexi
Also do not value ecosystems and are heavily influenced by the political economy
The food and non-food supply chains below comprise the sub-nexi, and provide a useful framework to understand the influence of the political economy on the Nexus and potential leverage points for valuing ecosystems.</td></tr>
<tr><td>The Water/Food/Trade Sub-Nexus, best understood as the Food Supply Chain</td><td>The Water and Sewage Services Sub-Nexus, best understood as the WASH Supply Chain</td><td>The Energy/Carbon/ Climate Change Sub-Nexus, best understood as Energy Supply Chains</td></tr>
</table>

Figure 5.1 The organizational architecture of the Nexus and sub-nexi. Source: Authors.

As mentioned above, there are important distinctions between how food and non-food supply chains, identified as sub-nexi above, impact on nature's ecosystems. Figure 5.2 is an attempt to quantify approximately the proportions of water and energy consumed globally in the food and fibre supply chains and in non-food supply chains. The evident asymmetries in the consumption of water and energy in the food/fibre supply chains and in the non-food/fibre supply chains are indicators of their respective impacts on nature's stores of non-renewable water and energy resources, and on renewable water and energy systems and ecosystems.

In Figure 5.2, levels of consumption are determined to some extent by technology and levels of socio-economic development. Therefore, levels of consumption differ in OECD, BRICS and developing economies. In food supply chains everywhere, ecosystem impacts are mainly determined by farmers who manage and mis-manage almost all of the 92 percent of our water consumption that is embedded in food supply chains. Consumer choice is also very important. For example, unhealthy diets waste natural resources, as does the non-consumption of purchased food – for example, in OECD economies where at least 30 percent of purchased food is thrown away. Energy supply chains in manufacturing and transport also impact ecosystems. Energy supply chains providing energy for domestic users at home have ecosystem impacts

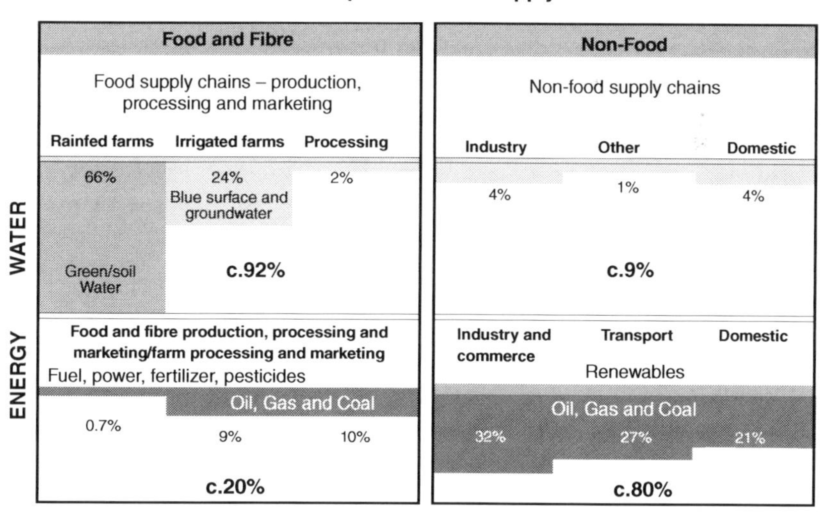

Figure 5.2 Global water and energy consumption providing goods and services in private sector supply chains. Source: Allan (2014), based on own unpublished research, and Water Footprint Network. Energy use based on Williams and Simmons (2013). Water and energy consumption in food and fibre supply chains and in non-food supply chains, showing the extreme asymmetries in levels of water and energy consumption.

Note: there are inevitable and unquantifiable error bars associated with such global estimates.

determined by technology as well as by wise and unwise consumer choices. The role of domestic consumption of energy is very significant and is equivalent to the total energy consumption in food and fibre production and distribution.

This section has provided evidence that natural ecosystems are impacted differently in the food/fibre sub-nexus and in the non-food sub-nexus because of the different proportions of nature's water and energy consumed in the respective sub-nexi. The next section will take the analysis further by identifying additional differences between the sub-nexi in the very important area of costing (or not costing) natural resources inputs and externalities.

The extent to which ecosystems are valued in Nexus and Sub-Nexus systems – consequences

The analysis so far has highlighted the hierarchical nature of a poorly understood Nexus system and the much better understood sub-nexi, which can be identified and understood as food/fibre and non-food/fibre supply chains. We are far from having an adequate understanding of the operational mutuality of the major supply chains (or sub-nexi) in our market economies, and from capturing the distorting processes that negatively impact ecosystems.

However, a number of things are clear. First, in the food/fibre sub-nexus that accounts for 92 percent of global water, farmers, who manage almost all of this water, manage consumption water ecosystems. Second, water is almost everywhere assumed to be a free or very cheap good in these supply chains. Third, in the trading, processing, marketing and consuming parts of the food/fibre supply chains, water is valued as an input. In all the non-farm elements of the supply chain, blue water is costed, albeit inadequately.

As mentioned above, the private sector often has to address Nexus issues and ecosystem degradation associated with the provision of water and energy. Water and/or energy are over-consumed in their operations. One of the reasons that the hotspots approach has dominated research on and analysis of the Nexus is because the private sector players engaged in food and energy supply chains have encountered serious operational problems and/or have identified significant current and future risks.

For example, in too many of the world's food bowls and centres of food and fibre production, farmers have encountered crises where irrigation is a key mode of water provision. In the United States, California has been living on borrowed time for decades with respect to both surface and groundwater supplies. Farmers over the Ogallala aquifer in Texas and in the states to the north are using that aquifer unsustainably. Wherever we have irrigated we have run out of water. Farmers in the Aral Sea Basin countries and in all Middle Eastern and North African countries, as well as those in the Punjab, the Yellow River and the North China Plain and Australia, have all placed unsustainable demands on their blue water resources. The bioethanol industry is another example where natural resource consumption contradicts sustainable long-term outcomes. In producing this cleaner source of energy, a great deal of water is devoted to the production of sugar and other field crops. Only Brazil has abundant green and blue water resources that make the production of ethanol appropriate. Where bio-ethanol is being produced elsewhere, the practice contradicts Nexus principles. These principles would guide supply chain operators to avoid the use of water in ways that have negative environmental impacts or would be sub-optimal economically.

It is not only farmers who encounter difficulties in marshalling water and energy inputs sustainably. Mining industries and especially the oil and gas industries also encounter serious operational problems. Sometimes they do not have enough water. On other occasions their operations pollute water – for example, in oil extraction from Canadian tar sands and in fracking.

What all these cases have in common is that they are all supposedly commercial processes, run by managers with significant interests in remedying problems with water and/or energy scarcity once such problems emerge. As we have shown above, it was corporate players who pushed the Nexus to prominence through the World Economic Forum. This push was a recognition that Nexus issues have potentially catastrophic consequences for businesses. Among these corporate players and the numerous actors that operate in the private sector food and non-food supply chains, the concept of the Nexus is most eagerly grasped. There tends to be a focus on such hotspots instead of addressing the main issue, which is that ecosystems and water are neither valued nor effectively accounted for.

Leverage points in the Nexus that could promote ecosystem stewardship

It has been shown that three major supply chains – first, food and fibre; second, water and sewage services; and third, energy services – are major components in any Nexus where theorists and practitioners have ambitions to bring about sustainable systems of provision (see Figure 5.1). Existing food and fibre supply chains and our energy supply chains are unsustainable (cf. Tilman *et al.*, 2001; Foley *et al.*, 2005; Godfray and Garnett, 2014).

The water and sewage services supply chains need further investment and development. With such investment they could potentially be sustainable as their demands for water and energy can be met sustainably even for the higher populations of the future. However, our energy supply chains are unsustainable because of their impacts on the atmosphere. Sourcing energy is different from sourcing water. Two renewable water sources, blue and green water, account for the water consumed in our economies. Energy is different. We can mobilize at least six non-renewable energy resources. Coal, oil and gas are the major non-renewables. There are also at least six renewable energy sources: hydropower, wind, solar, tidal, wood and bio.

Water and energy are therefore different with respect to substitution. With water, there can only be discussion on how to deliver and finance the augmentation of green water with blue water. The history of energy consumption on the other hand has been a constant story of contestation. Wood was replaced by coal. Oil and gas have become the main providers of energy and have progressively substituted for coal, especially in transportation but also in most industrial and domestic energy consumption.

With water the potential leverage could be by those who produce, trade and market food and fibre in supply chains. Food and fibre supply chains consume the majority of the water, and most of this water is managed on farms. Put this way, it is obvious that farmers are the key players. They determine levels of water consumption and they steward, or not, water ecosystems on behalf of society. However, they can only do this if they have livelihoods and incentives that make it possible to contribute stewardship services. The leverage that would help farmers manage water better should be exerted via supply chains that price food to recognize the value of natural resources. The politics and awareness required to install such a system are so challenging that this type of leverage is unlikely to be adopted unless there is a crisis that impacts both the rich and the poor. Environmental economic indicators point one way, while the political imperatives of ensuring the availability of cheap food point the other. It is currently impossible to leverage change via the existing market economy.

In the absence of the adoption of a market system that would change farmer behaviour, it will be necessary for governments and consumers to leverage change. Governments could formalize voluntary systems of sustainable practice as well as the associated voluntary certification systems. Ideally, the private sector traders, food processers and supermarkets would then align with such trends. However, the cheap food mantra is too strong. The main way that farmers will be leveraged to adopt sustainable practices will be by focusing government-financed farm payments. These will have to be directed at farmers who can make the biggest contribution to sustainable water management and water ecosystem protection.

Leverage in the energy sector will be very different. In contrast to sourcing water, there are not just two main sources but many renewable options that are already enabling the substitution of clean energy sources for dirty ones, and will do so increasingly in the future. Clean targets have already been identified and there are strong voices advocating sustainable energy policies in society, civil movements, the private sector and governments. However, these have not yet reached a tipping point, and to date they have not been strong enough to counter the vested interests that continue to produce dirty energy. As a consequence, the uncertainty of political commitment has determined a slow uptake of sustainable practices.

Energy supply chains, like food supply chains but not to the same extent, are mainly in the private sector. Energy markets are also seriously distorted by subsidies, although again not to the same extent as the mis-pricing of water in food supply chains, where water is regarded as an essentially free good on farms. In these circumstances, investors are beginning to play a major role by divesting from coal and promoting cleaner energy supply chains that do not have such heavy impacts on climate change and ecosystems. KPMG, a private sector player, has reported on sustainability integration in European banks with the support of the international NGO WWF, and has shown that these banks are increasingly investing responsibly in energy supply chains (KPMG, 2014). Other organizations are also following suit. For example, Stanford University recently announced that it will divest coal stocks from its $18.7bn endowment (Stanford News, 2014). Such divestment and additional investment in clean solutions needs to reach the $2.8tn cited as the possible level of such investment by 2024. However, the movement towards more sustainable energy supply chains clearly demonstrates that positive change is emerging, albeit slowly.

Conclusions

When water and energy are not sustainably managed, there are significant ecosystem and environmental consequences. Food and non-food supply chains are at the heart of the Nexus because they account for most of the water and energy consumed globally. The ways these supply chains operate, and the rules within this sub-nexi structure, helps us understand and research the Nexus. To understand the Nexus it is necessary also to take into account political and market forces in the form of subsidies, profit seeking and state agendas. These forces powerfully influence the value attributed to, or not attributed to, ecosystem services and the environment on which they depend.

A Nexus approach that uses supply chains as its framework highlights the importance of getting players in the food, energy and water supply to change their modes of operation and value ecosystems. Scientists and professionals are beginning to identify sound data on ecosystem limits (cf. Steffen *et al.*, 2015), but they have not been able to provide an agreed system structure and associated metrics to clarify the measures that should be taken to remedy existing problems and anticipate future problems.

Businesses and corporations tend to approach the Nexus from a hotspot approach that looks at inputs and outputs in supply chains at local scales in particular geographies. This approach is insufficient and does not address the wider forces in the political economy that dis-incentivize the stewardship of ecosystems. Significant changes have to be made in the ways that food, energy and water supply chain operators value natural resource inputs and account for ecosystem impacts. Without these changes, our food and energy will continue to be managed and produced in fragile and unsustainable systems that are vulnerable to increasingly devastating shocks, which will be further exacerbated by climate change. In addition, the Nexus approach will fall short of its goals of putting in place sustainable practices for the use of natural resources. In order for the Nexus approach to be sustainable, it must understand and engage with the very highly politicized supply chains that determine how water and energy are consumed.

Notes

1 There are no reliable global, regional or national data on the relative proportions of subsistence, family and corporate farmers or of the relative productivity of these farming modes.
2 The availability of green water or effective rainfall is to some extent reflected in land values, but the value of green water is not identified and captured in current accounting and reporting systems.

References

Allan, J.A., Keulertz, M. and Woertz, E. (2015) 'The water-food-energy nexus: An introduction to nexus concepts and some conceptual and operational problems', *Water Resources Development*, 31(5), 301–311.

Allouche, J., Middleton, C. and Gyawali, D. (2015) 'Technical veil, hidden politics: Interrogating the power linkages behind the nexus', *Water Alternatives*, 8(1), 610–626.

Berry, G.R. (1974) 'The oil lobby and the energy crisis', *Canadian Public Administration*, 17(4), 600–635.

Bloomberg (2014) *Fossil fuel divestment a $5 trillion challenge*, White Paper, Bloomberg New Energy Finance. Available online at: http://about.bnef.com/white-papers/fossil-fuel-divestment-5-trillion-challenge/ at 23 July 2014 (accessed on 20 July 2015).

Carrington, D. (2015) Fossil fuels subsidised by $10m a minute, says IMF. *The Guardian*. Available online at: www.theguardian.com/environment/2015/may/18/fossil-fuel-companies-getting-10m-a-minute-in-subsidies-says-imf (accessed on 22 July 2015).

Coady, D., Parry, I., Sears, L. and Shang, B. (2015) 'How large are global energy subsidies?' *International Monetary Fund*. Available online at: www.imf.org/external/pubs/cat/longres.aspx?sk=42940.0 (accessed on 7 July 2015).

Dupar, M. and Oates, N. (2012) *Getting to grips with the water-energy-food 'nexus'*. Climate and Development Knowledge Network (London). Available online at: http://cdkn.org/2012/04/getting-to-grips-with-the-water-energy-food-nexus/ (accessed on 10 July 2015).

Durant, R.F. and Holmes, M.D. (1985) 'Thou shalt not covet thy neighbor's water: The Rio Grande Basin regulatory experience', *Public Administration Review*, 821–831.

Foley, J.A. *et al.* (2005) 'Global Consequences of Land Use', *Science*, 309, 570–574.

Gleick, P.H. (1994) 'Water and energy', *Annual Review of Energy and Environment*, 19, 267–299.

Godfray, H. C. J. and Garnett, T. (2014) 'Food security and sustainable intensification', *Phil. Trans. R. Soc. B*, 369, 20120273. Available online at: http://dx.doi.org/10.1098/rstb.2012.0273

Goodman, D., Redclift, M., Marsden, T. and Little, J. (1990) 'The farm crisis and the food system: Some reflections on the new agenda', *Political, Social and Economic Perspectives on the International Food System*, 19–35.

Handfield, R.B. and Nichols, E.L. (1999) *Introduction to supply chain management* (Vol. 999). Upper Saddle River, NJ: Prentice Hall.

Hoogeveen, J. (2015), quoted in Allan, J.A., Keulertz, M. and Woertz, E., 'The water, food, energy nexus concepts and some conceptual and operational problems', *International Journal of Water Resources Development*, (31(3), 301–311.

Ingram, H.M., Mann, D.E., Weatherford, G.D. and Cortner, H.J. (1984) 'Guidelines for improved institutional analysis in water resources planning', *Water Resources Research*, 20(3), 323–334.

KPMG (2015) *Ready or not: An assessment of sustainability integration in the European banking sector*, KPMG.

Middleton, C. and Allen, S. (2014) *The (Re)discovery of "the Nexus": Political economies and dynamic sustainabilities of water, energy and food security in Southeast Asia.* Asia Pacific Sociological Association Conference Paper. Available online at: http://carlmiddleton.net/wp-content/uploads/2014/12/MiddletonAllan_Nexus_FINAL.pdf (accessed on 11 July 2015).

Scott, C.A., Kurian, M. and Wescoat, J.L. (2014) 'The water-energy-food nexus: Enhancing adaptive capacity to complex global challenges', in M. Kurian and R. Ardakanian (Eds), *Governing the nexus: Water, soil and waste resources considering global change*, Dordrecht: Springer, 15–38.

Stanford News (2013) *Stanford to divest from coal companies.* Available online at: http://news.stanford.edu/news/2014/may/divest-coal-trustees-050714.html

Steffen, W., Richardson, K., Rockström, J., Cornell, S.E., Fetzer, I., Bennett, E.M., Biggs, R., Carpenter, S.R., de Vries, W., de Wit, C.A., Folke, C., Gerten, D., Heinke, J., Mace, G.M., Persson, L.M., Ramanathan, V., Reyers, B. and Sörlin, S. (2015) 'Planetary boundaries: Guiding human development on a changing planet'. *Science*, 347, 736–747.

Tilman, D., Fargione, J., Wolff, B., D'Antonio, C., Dobson, A., Howarth, R., Schindler, D., Schlesinger, W.H., Simberloff, D. and Swackhamer, D. (2001) 'Forecasting agricultural driven global environmental change', *Science*, 292, 281–284.

Verhoeven, H. (2015, forthcoming) 'The nexus as a political commodity: Agricultural development, water policy and elite rivalry in Egypt. *International Journal of Water Resources Development.*

Williams, E.D. and Simmons, E.J. (2013) *Water in the energy industry: An introduction*, London: BP International. Available online at: http://www.bp.com/content/dam/bp/pdf/sustainability/group-reports/BP-ESC-water-handbook.pdf

World Economic Forum (WEF) (2011) *Water security: The water-food-energy-climate nexus.* Washington, DC: Island Press.

Part 2

Urban challenges of the Nexus

Local and global perspectives

6 The contribution of innovation in urban resilience and sustainability to realizing the urban Nexus

Nicholas You

Introduction

Future-proofing our cities and metro-regions is becoming a major preoccupation for all tiers of government, the communities that live in those cities, and the private and commercial sectors that invest in them. Not a day goes by without news coverage of a city or metro-region being flooded, affected by a landslide, or hit by a heatwave, drought or strong winds. Many of these events wreak substantial damage to lives, property and infrastructure and increase the transaction costs for all forms of commerce and industry.

The urban Nexus (water–energy–food and the impacts of climate change) has become all the more relevant in light of these challenges. Many of the urban systems that are failing or are prone to failure were designed in splendid isolation from one another. Typical examples include water supply and sewerage, zoning and storm water drainage, energy supply and conservation, and so on. In almost all cases, making our cities more resilient requires that we revisit how these systems are conceived and run. It also requires that we revisit the roles and responsibilities of different stakeholders and the over-arching governance framework. Finally, but importantly, rethinking our systems begs the fundamental question of who we are making cities more resilient for. In the final analysis, the urban Nexus cannot be just about energy, water and food – it also has to be about social equality and justice.

The purpose of this chapter is to look at lessons learned from a unique snapshot of case studies in urban innovation from different regions and contexts. The case studies share one thing in common: they were peer reviewed and selected by an independent panel of experts from more than two hundred initiatives submitted for the 2014 Guangzhou International Award for Urban Innovation.[1] The thematic areas of the finalists for 2014 very much reflect the real-world challenges facing cities today. Unambiguously topping the list, and as the urban world's premier concern, are nine cities that are seeking ways and means to protect themselves and others against the ravages of climate change. The purpose of this chapter is to provide a preliminary overview of some of these city initiatives as a contribution to further analysis and discussion of how the urban Nexus can help make our cities and metro-regions more resilient and sustainable.

Local governments preparing to meet future challenges

The city of Gwangju in South Korea has taken a very innovative approach to climate change mitigation. Five years ago it initiated a programme to engage citizens in voluntary carbon-saving steps. Gwangju's Carbon Bank System currently involves 330,000 households, representing 62 percent of the city's population of 1.48 million. The Carbon Bank System calculates reduced amounts of carbon dioxide through voluntary energy and water-saving efforts by households and turns them into points. The Carbon Bank involves the active participation of the Korea Electric Power Corporation, the Gwangju Metropolitan Waterworks Authority and Hae Yang City Gas. All three corporations provide data to the system based on participating households' consumption. Gwangju City, with the assistance of a commercial bank, calculates the reduction in greenhouse gas emissions and turns the amount into points which are credited to the electricity, gas and water bills of the participating households.

The Gwangju Carbon Bank System provides an interesting case of two aspects of the urban Nexus:

1 how to provide citizens with a tangible means of seeing their actions reflected in terms of CO_2 reductions (the common good) and in their utility bills (personal benefit)
2 tying together water, electricity and gas.

This bridging of "silos" is perhaps one of the most important hurdles to be overcome when striving for low-carbon urban management and development. Through this integrated system, the city can analyse and evaluate how different actions in different areas affect the total equation of greenhouse gas (GHG) emissions and identify which areas need greater focus to continue to progress towards lowering the carbon footprint of the city. Since the inception of the programme in 2008 GHG emissions have decreased each year, most recently by 135,000 tons.

In one of the world's most rapidly growing cities, a new government-mandated programme – Estidama (Arabic for "sustainable") – aims at making Abu Dhabi in the UAE more environmentally responsible and sustainable for its 920,000 inhabitants. The UAE has witnessed unprecedented urban development and poor resource management, which has resulted in one of the world's largest carbon footprints per capita. A key component of Estidama is the Pearl Rating System (PRS), a sustainability rating system introduced in 2010. The PRS guides projects through design, implementation and management. Although sustainability-rating systems exist elsewhere, Estidama is the only such programme designed and implemented in the Middle East. It benefited from different stakeholder inputs and is designed for arid regions. Estidama is at the technological edge of green building monitoring systems, adapting the most effective existing systems to local conditions and context. The programme is free and has a training component to raise awareness and capacity within the construction industry. The programme targets energy use reduction of 31 percent, water use savings of 37 percent, and 65 percent of construction waste diverted from landfill. There is a mandatory audit procedure for each project. The rules ran into initial resistance from industry groups that feared increased costs and more difficult project approval. However, independent analysis has confirmed that cost increases are negligible. Less tangible but no less important is the improvement in living and working conditions as new constructions require environmentally sound and healthier building materials, less aggressive cooling systems and better and more user-friendly designs.

The Estidama system appears to be very similar to other green building and design systems that seek to reduce the ecological footprint of the built environment. Nonetheless it embodies some important considerations that are particularly relevant to countries and cities undergoing very rapid urbanization and accelerated economic growth. These considerations are the need to identify and engage with the most relevant actors. Three unique challenges that are associated with rapid urbanization and economic growth include:

1. the rapid replacement and renewal of the built environment, sometimes as frequently as every ten years
2. rapid turnaround in the ownership and tenancy of real estate
3. rapid increase in real estate value and prices.

These challenges have one particular effect: the lack of incentive for developers, contractors, owners and users to go green. These challenges led Abu Dhabi to target the construction and building management industries as the key actors and entry points for implementing change. Further steps will require long-term efforts to educate inhabitants that come from very diverse social and cultural backgrounds, who have widely differing perceptions of sustainability.

Box 6.1 The fifteen short-listed city initiatives

Abu Dhabi, UAE: tailor-made cutting edge green building system

Antioquia, Colombia: knitting together regional territory through innovation in education

Boston, USA: empowering youth through participatory budgeting

Bristol, UK: citizen-centric approach to the Smart City

Buenos Aires, Argentina: dialogue for decision making for urban projects

Christchurch, New Zealand: community-led transitional city for post-disaster reconstruction

Dakar, Senegal: city-driven innovation for municipal financing

Eskisehir, Turkey: Eskisehir City Memory Museum

Gwangju, South Korea: incentivizing households to reduce GHG emissions through devolved targets

Hamburg, Germany: socially-inclusive approach to building a zero-carbon district

Hangzhou, China: innovation in large-scale operation and maintenance of public service delivery

Jakarta, Indonesia: engagement of political leadership in pro-poor participatory process

Linköping, Sweden: long-term consensus-driven alignment for attaining carbon neutrality

Melbourne, Australia: landscaping approach to cooling the city

Rio de Janeiro, Brazil: using big data for integrated risk management and action

Source: www. guangzhouaward.org

Linköping, Sweden has a bold goal and an equally bold approach to becoming a carbon neutral city by 2015. The city council's road to that goal was launched through broad-based collaboration and partnership with residents, employers, universities, other cities and national and international networks. The municipality has also sought to lead by example: it uses renewable fuels (over half its vehicles use biogas); it specifies climate criteria in its procurement processes; it regularly communicates climate and environmental issues to its employees and residents; and it works closely with Linköping University to develop methods and technologies to reduce CO_2 emissions and establish a Biogas Research Centre. Two new combined heat and power plants have been built, with 95 percent of homes and most of the city's 150,000 residents already connected. City buses are fuelled by biogas produced from livestock manure and food waste. The results are already emerging: CO_2 emissions are down by 25 percent since 1990, energy consumption in schools and hospitals have been reduced by 5 percent and are on a trend for further reductions.

In many ways Linköping is a good illustration of how a city can implement a truly integrated approach called for by the urban Nexus. The local authority fulfills a leadership role in setting goals and targets, mobilizing stakeholders and working across its own administrative silos. Integrating climate criteria into its procurement processes epitomizes this leadership role, because in many cities of the size of Linköping the local authority is one of the largest procurement agents. Changing procurement policies can thus have a profound effect on the behaviour of industry and service providers beyond Linköping's own borders. Another innovative aspect of Linköping's initiative is the use of animal and food waste to produce biogas – an original approach in a climate where producing local food is difficult for a significant part of the year.

The 6.3 million inhabitants of Rio de Janeiro have been hit hard by repeated Atlantic storms imperiling the city. This especially affects the mostly low-income settlements that are located on the high slopes surrounding the metropolis, which are prone to devastating landslides. Following a vicious storm in 2010 Rio de Janeiro decided to create a centre that operates twenty-four hours a day, staffed by officials from thirty city departments. This centre has become a global model showing the benefits that can be derived from collaboration, alignment and data-sharing across city divisions. Since the facility went online, employing some of the latest information communication technology and weather forecasting systems, there have been no deaths caused by landslides. The model has subsequently had many other benefits for the day-to-day management of the city. Traffic emergency response time has been reduced significantly, with citizens alerted about traffic snarl-ups and accidents and redirected to the best routes. Data gathered for the centre also enables the identification of neighbourhoods with higher dengue fever infection rates. In planning the facility, Rio officials visited alert centres in Madrid, Seoul and New York, and have since forged cooperation with Johannesburg as it plans a similar system.

Rio de Janeiro's take on big data analytics and sharing constitutes an important lesson for many cities in both the north and the global south. Data analytics needs to be issue-driven rather than technology- or data-driven. In the case of Rio, the initial challenge was very clearly defined: reduce deaths and injuries due to the heavy rain that often causes landslides, especially for the most vulnerable segments of the population living in precariously located favelas. This led to the sharing of data and knowledge across departments that had previously ignored each other's existence, such as land-use planning and meteorology, soil and sub-soil mapping, and informal settlements. Achieving this narrowly focused objective led to other "discoveries" about how data sharing and analytics could be applied to very different issues and sectors. However, the key to Rio's success lies in one of the fundamental drivers of the urban Nexus approach, namely the identification of context-specific objectives that in turn define the scope of integration required and the actors that need to be involved.

Smart City Bristol is a collaborative programme between the public sector, business and the community. The main aim is to use smart technologies to help meet the ambitious city target to reduce CO_2 emissions by 40 percent by 2020 from a 2005 baseline, as well as other social and economic objectives.

The programme was launched in 2011, focusing on Smart Energy, Smart Transport and Smart Data. It includes pilot projects such as smart metering, smart grid, electric vehicles and open data, alongside permanent initiatives such as a Traffic Control Centre and a Freight Consolidation Centre. A distinctive trait of Smart City Bristol lies in its collaboration with micro-electronic, environmental technology and creative/digital companies who are working with communities to make them smart. One such example is Bristol's Living Lab in Knowles West, which groups people who are actively involved in the creation and evaluation of technologies which they will ultimately use.

The projects are applied within the Bristolian context using their own approach to smart cities. Smart City Bristol is about putting its more than 430,000 people, rather than technology, at the heart of a smart city. It is how people interact with technology that helps to inform behaviour change and help the city achieve its aims. Smart City Bristol is also evolutionary in that it builds upon the lessons learned from previous projects to inform new ones.

The current accomplishments include:

- Securing an additional £7m in funding for Green Capital activity
- Becoming one of the inaugural 100 Resilient Cities, a pioneering initiative funded by the Rockefeller Foundation
- Achieving the 20 percent energy reduction target as part of 3e Houses
- Decrease in energy consumption by schools and council offices as part of the "smart spaces" project
- Increased awareness and engagement with climate change issues via smart city projects in general

Bristol is already the most energy and waste-efficient major UK city, and it is planning to meet future needs by managing resources even more efficiently. Smart City Bristol has contributed to the image of Bristol as a centre for innovation and a leader in the adoption of "smart" solutions to city-level problems.

Bristol's smart city approach shares a common feature with Rio's approach to data sharing and analytics: the process is context- and people-driven, with the inhabitants of the city having a major say in the choice of technological solutions. This has two major implications, which are critical to the successful implementation of the urban Nexus. The first is that the issues and challenges are fully understood by all. The second is that the costs and benefits of the proposed solutions, including sophisticated technological solutions, are also fully understood by the public. This commitment goes beyond conventional participation and token engagement – it involves the continuous education of the citizens to involve them in the strategic and operational aspects of making cities and communities more resilient and sustainable.

Integrating people's needs with environmental goals

In 2005 the city of Hamburg decided to embark on a socially inclusive approach to zero-carbon urban development. A pilot project was established for the redevelopment of the Wilhelmsburg neighbourhood of the city by hosting the International Building Exhibition (IBA). An IBA Partnership was established which brought together private companies and the local community. As a result over seventy projects were developed around three themes, including cities and climate change. Wilhelmsburg has 55,000 inhabitants living on an island in the Elbe. It is particularly vulnerable to flooding. It is also an ethnically diverse and low-income community with a physical environment affected by industrial and transport infrastructure. The projects are based on maximizing the use of local energy resources such as energy savings and energy efficiency, and using the technological opportunity as a means of strengthening local economy development. The aim is to achieve 100 percent local renewable supply by 2025 and 100 percent renewable heat by 2050, making the Elbe island carbon neutral. The IBA is also unique in its governance structure as it is incorporated as a limited liability company. It thus has a certain amount of independence from classic administrative hierarchies and can act more like a private enterprise.

The IBA Partnership comprised about 150 companies and institutions as well as the local inhabitants, involving them in numerous workshops and fora. Less bureaucracy and political interference allowed the IBA to set ambitious targets and to mobilize partners. Already scheduled projects will ensure that 54 percent of heat production and 14 percent of the overall energy demand will be renewably produced by the end of 2015. The IBA Hamburg Model and Wilhelmsburg's Climate Protection Renewable strategy has made the city of Hamburg one of the frontrunners when it comes to socially inclusive and environmentally sound urban development. Additionally, the IBA is sharing the knowledge generated with other partner cities and is currently scaling up the Wilhelmsburg pilot to other parts of the city.

The Wilhelmsburg project underscores one of the major challenges of the urban Nexus approach – that of overcoming administrative silos and bureaucracy spanning several electoral mandates. Despite being a recognized leader in many aspects of urban sustainability and resilience, the city of Hamburg opted in this case for an outsourced approach to ensure policy continuity over time. This has helped to de-politicize the project. Equally importantly, it has enabled the project to be managed and financed on a multi-year basis rather than according to the annual budgetary cycle of most municipal authorities. This in turn helps mitigate the risk for the private sector to engage in socially inclusive development.

Jakarta is a mega-city of 9.6 million people that lies in a delta of thirteen rivers, with 40 percent of its land lying below sea level. Not surprisingly, it faces a huge crisis of flooding, algae and water pollution. Its Pluit Reservoir Revitalization Project represents the city's significant effort to improve water storage capacity, reduce urban flooding and improve the quality of its prime water source. It is predicted that 80 percent of North Jakarta will be 5 metres below sea level by 2030. As a response, the Pluit Reservoir Revitalization Project has been introduced within the framework of the Jakarta Water Management Plan for 2030. Covering a catchment area of 2,083 hectares in

North Jakarta, the reservoir will play a major role in mitigating the annual cycle of flooding. However, this requires a housing solution for many of the households who are currently living on the banks of the reservoir. Improving storage capacity of the reservoir requires the first phase relocation of 3,000 squatters around the reservoir's banks, and transforming adjacent areas into parks and high-quality public open space. The main challenge was to relocate the squatters without violating their human rights. This was ensured by strong leadership. The Governor himself contacted the people directly, listened to them and communicated with them frequently. At the same time the Governor encouraged private sector partners to donate attractive furniture for the new housing, and explained to the local people the importance of maintaining the reservoir area. The city's water management plan, envisioned for roll out from the present through to 2030, is seen as a way to address climate change in a socially conscious way. It includes government partnerships with a corporate sector that is expected to benefit – along with the public – from new and less threatened property development.

The Pluit Reservoir Revitalization Project is yet another illustration of how to work together with the private sector in realizing the urban Nexus. As in the case of Wilhelmsburg in Hamburg, the local authority engaged with private real estate developers to find a win–win solution. On the one hand, the resettlement of squatters along the lakeshores represents an exercise in adaptation that will prevent a predictable loss of lives and property as the risks of flooding increase over time. At the same time, the restoration of a pristine lake environment creates value for high-end real estate development, in exchange for a cross-subsidy for the relocation and re-housing of the squatters.

Between 1995 and 2009 the city of Melbourne suffered extreme hot weather resulting in severe drought, water shortage and heatwaves that killed several hundred people out of a total population of just over four million. The immediate response of the city was to plan a 90 percent reduction in potable water use. This included cutting irrigation support to the city's urban forests and a plan to remove 40 percent of the city's trees. Ironically, this solution underestimated the value of green spaces and ecosystem support, which are critical to climate change adaptation and mitigation. Realizing the need for a strategic long-term strategy, in 2010 the city appointed a new Urban Landscape Team. The team produced the open space strategy and the urban forest strategy. Since 2010, $40 million have been invested in related initiatives including urban forests and shrubbery, green space and rainwater harvesting, permeable paving, and protection of waterways and wetlands. The goal is to cool the city by 4 degrees Celsius and thus lower energy use for cooling. Fifteen thousand trees have been planted, forty streets retrofitted to improve permeability, and one of the world's first in-road storm water harvesting systems started. A four-year citizen's engagement programme educates and mobilizes citizens. The city of Melbourne provides the bulk of the funding while the regional and federal governments have also contributed. Other partners include the universities of Melbourne and Victoria for related research, and the media for public awareness.

Rebuilding better after devastating disasters has been a strong motto that has rarely been realized in practice. Christchurch's approach is, in the words of one of its councilors, the "new normal." This philosophy guides virtually all of Christchurch's policies, strategies and decisions in post-disaster reconstruction, spanning from maximizing benefits through well-coordinated investments in reconstruction, avoiding duplication of efforts in service provision, infrastructure and facilities, and not least, how to restore quality of life not just in material terms but through the restoration of a sense of place, local identity and communal solidarity. A question here is why it seems to take events of disastrous magnitude to bring common sense into urban planning and management.

With 80 percent of its eight million residents and commuters identifying a serious traffic problem in the city, Hangzhou launched China's first public bicycle project. Currently serving some 280,000 passengers daily, the system (free for the first hour) complements the city's extensive bus system. Run by the newly-formed Hangzhou Public Bicycle Development Company, it represents a model of government-led enterprise, claimed to be the world's largest bike-sharing programme that doesn't require government funding beyond initial capital. As well as fees for bike use (imposed after one hour), the company raises significant funds by selling advertising space on the bike docking station kiosks, which are designed as mini-convenience stores selling everything from food and beverages to mobile phone payment and credit facilities. Three key features of the Hangzhou initiative are:

1. partnerships with universities, to monitor trends and issues in the use of the bikes
2. a unique management system designed to overcome the most frequent problem areas of bike-sharing systems: service points and getting bikes to where they are needed when they are needed
3. a responsive hotline support system including attendants at the busiest points to immediately attend to customer needs and implement users' recommendations for the continuous improvement of the system.

Hangzhou's bike-sharing business model is not unique in and of itself. What is unique is its scale of operations, serving more than a quarter of a million riders per day, and its payback model. The single biggest contributing factor to its success, in the words of its managing director, is the fact that docking stations are on average only 300 metres apart, adjacent to bus stations and in close vicinity to metro stations. This allows the kiosks to cater for much more than just the users of the bike-sharing system, reaching out to bus and metro riders as well. This is a perfect example of how the successful implementation of an urban Nexus solution often depends on lateral thinking, connecting dots that have remained unconnected for decades.

Some preliminary observations

The nine initiatives described above provide only a small snapshot of what cities and regions are doing globally to cope with and mitigate the effects of climate change. Nonetheless, they represent a wide variety of entry points and approaches. The first lesson to be learned is that there is no one size that fits all when it comes to climate change adaptation and mitigation. However, this said, there do appear to be several common threads that could help accelerate mutual learning and exchange. The first of these is the urgent need to question the old methods of doing things – a fundamental premise of the urban Nexus approach. This is perhaps most evident in the case of Melbourne, where the intense suffering from extreme heat and drought triggered an initial response of saving water at all costs, including the removal of trees and forests that required irrigation. Fortunately, this knee-jerk reaction was replaced with more strategic and long-term thinking that led to a simple yet compelling objective of reducing the urban heat sink. This requires the expansion rather than the contraction of green space, but the end result is a lower carbon footprint, lower energy consumption and lower water consumption.

A completely different entry point is demonstrated in the case of Hangzhou, where the main challenge viewed by its inhabitants is not climate change adaptation or mitigation but traffic congestion. In trying to address what the citizens viewed as a priority issue, the local authority came up with a response that is not only effective in terms of providing an alternative mobility option, but also in helping to reduce the carbon footprint of urban mobility. Its innovative dimension resides in its client orientation, which has allowed the operator to solve many common problems that plague bike-sharing schemes in many places around the world. This bike-sharing programme that is inspiring many other cities in China may well mark an ironic turning point in bicycle use in China's recent history – which went from being the country with the most bicycle users thirty years ago to the country with the most cars and traffic jams today. Hopefully this trend to reaffirm the bicycle as an important mode of urban mobility will receive great impetus from the exemplary manner in which Hangzhou has managed its bike-sharing programme.

Abu Dhabi, Bristol and Gwangju, despite their radically different approaches, share a common urban Nexus objective – how best to change human behaviour. In the case of Abu Dhabi, the Pearl Rating System uses a combination of compliance, training, awareness building and objective evidence to overcome initial resistance to change on behalf of the construction industry. Changing attitudes and behaviour on the supply side is of critical importance for cities, which are growing rapidly and where a significant portion of the population consists of migrants that have very different views, attitudes and proprieties when it comes to sustainability. On the other hand, Gwangju combines awareness-raising at the household level with a tangible link between individual actions to save energy and water with the collective good of reduced carbon emissions, making this link tangible as an individual benefit through reduced utility bills. Making the link visible and tangible is also part of Bristol's approach to the smart city. This approach puts the community in control of the technologies that will make a difference to their quality of life while achieving the goal of reducing carbon emissions. The premise of Bristol's approach is that if they are to change their behaviour, people must not only understand the innovative technologies that they will be using, but also be part of the process of evaluating and choosing them.

Jakarta and Hamburg both point to the importance of social inclusion in attaining bigger order and long-term sustainability goals. In the case of Jakarta, leadership was provided by a political leader – the Governor – who took it upon himself to engage with the urban poor to encourage them to accept an alternative housing solution both for the common good and to improve their living conditions and health. This win–win solution was careful to avoid any loss to the livelihood opportunities of the poor. It also incorporated benefits for real estate developers in the form of attractive parks and gardens to cross-subsidize the scheme. In the case of Hamburg, an alternative institutional and governance arrangement was devised to cut down on bureaucratic red tape and political interference, aligning private sector investment with creating better conditions for local economic development and job opportunities for a disadvantaged segment of the population. In both cases, alternative solutions were found that traditional approaches to urban development would have had great difficulty in implementing.

Linköping and Rio de Janeiro would seem at first glance to have little in common. However, these two initiatives both point to the importance of strategic vision and leadership. In the case of Rio, the initial preoccupation was to be able to provide sufficient advance warning of landslides affecting the favelas. The result of the endeavour led to a big data/smart city approach that is bringing benefits to a much broader cross-section of society, in the context of isues such as how best to avoid traffic congestion or timely preventative measures against tropical diseases. On the other hand, from the very start Linköping sought collaboration with multiple stakeholders, both domestic and international, in forging a compelling vision for a carbon neutral city. This singular goal has been translated into multiple initiatives that provide the basis for an integrated approach to improving quality of life and attaining environmental sustainability.

Summary

In summary, making cities more resilient and sustainable is not about technology, big data or smart cities per se; it is about a shared yet compelling vision, a set of ambitious goals, and a road map to addressing past ills and improving living conditions going forward. The contributory factors to success, in the case of the initiatives analyzed above, are all rooted in leadership through better governance, involving more participation and engagement as well as increased transparency and accountability. Herein lie the critical contributory success factors to realizing the urban Nexus.

Note

1 The goal of the award is to foster exchanges and learning between cities and regions around the world. Innovation, in the context of the Guangzhou Award, is understood to comprise new ideas, new policies and strategies, new technologies and business models, and new forms of management for tackling the social, economic and environmental pillars of sustainability and the cross-cutting domains of technology and good governance. The focus is very much on lessons learned from experience and the processes that underlie them.

7 Operationalizing the Urban Nexus

Increasing the productivity of cities and urbanized nations

Kathrine Brekke and Jeb Brugmann

Introduction

The increasing centrality of cities in the global economy is part and parcel of the global urbanization process. Both population and economic concentration in urban regions will continue until at least mid-century. As urbanizing nations and city-based corporations each seek to increase their productivity and pursue sustainable development, they will therefore need to find more and more of their solutions in the context of urban development and management.

Economists have reached a general consensus that there are many more contributing factors to economic productivity at both the firm and country levels than the traditional formulation of labour and capital productivity. Among other factors that have been statistically demonstrated to be key to productivity measurement and growth are: efficient use of material and energy inputs; employee and general population health; education and social welfare; technology and innovation; and urban agglomeration economies. The recognition of 'multi-factor productivity' (also known as total factor productivity) puts the focus of productivity growth beyond the borders of the individual firm. Many of the key productivity factors are determined within the urban regional context, where firms compete and collaborate, where the labour pool is educated, serviced, transported and cared for, where academic and research institutions and clusters drive innovation, and where urban infrastructures determine the efficiencies of energy, water and materials cycles.

Productivity is perhaps the single greatest common ground between those who primarily seek economic growth, those who primarily seek social development and those who primarily seek ecosystem health and the sustainable use and replenishment of natural resource endowments. This common interest in productivity has been highlighted, for instance, in the focus on eco-modernization over recent decades, of which the more recent focus on the 'circular economy' is part. The Nexus is a further area of study and practice within the eco-modernization field, focusing on the cross-optimization of multiple resource management systems in order to increase the total resource productivity – in terms of both efficiencies and resilience across these systems.

In reality, most of the current Nexus activity (i.e. in the water–energy–food/waste/soil Nexus) is inherently metropolitan in terms of the geography of Nexus opportunity, the sources of finance and operations (i.e. via city or metro governments and utilities), and in terms of governance. The overlap between eco-modernization efforts and efforts to increase productivity at the metropolitan regional scale therefore bespeak the need for an Urban Nexus approach. As outlined below, this approach can also be used to bring the social development agenda further into practical focus.

In 2014, GIZ and ICLEI-Local Governments for Sustainability published a report, *Operationalizing the Urban NEXUS: Towards resource-efficient and integrated cities and metropolitan regions*. This study represents the first concerted effort to apply the Nexus concept to cities and urban regions specifically, encouraging new modes of planning and action across institutions, policy arenas and asset operations at the scale of socio-economic organization that is becoming most dominant worldwide: the urban or metropolitan region. The remainder of this chapter condenses key points from the full GIZ-ICLEI report, which may be accessed as noted below.[1]

Why an Urban Nexus?

Continued, rapid urbanization is increasing demand for new or improved infrastructures, services and institutions that are capable of meeting a three-fold challenge:

1. providing larger urban populations with access to basic services and vital resources
2. sustaining economic development; and
3. managing resources within our planetary limitations while addressing the challenges of climate change adaptation and mitigation.

Attempts to satisfy the resource demands of growing urban areas and typical urban lifestyles has meant looking ever further afield for supplies – from metropolitan and rural hinterlands to regional and global scale sources. Meanwhile, the resource-efficiency and productivity of cities and urban regions is determined by the combined performance of the multiple jurisdictions, utilities and service departments, civic organizations, companies, and disciplines that share the responsibility for the city's social and economic development, urban form and infrastructures. The prevailing fragmentation of urban jurisdictions, services systems and urban management approaches has not enabled the optimization of cities across the totality of their highly interdependent systems. Conventional approaches to city building and management therefore overlook opportunities to increase multi-factor productivity (i.e. in terms of resource use, human factors and economic production), while also exposing growing regions to increasing risk.

Fragmentation in terms of physical design, governance, planning and investment results in myriad wasteful trade-offs, which are borne by industry, in public and private services operations and in household lives. So-called 'siloed' approaches and solutions aimed at just one sector or service area generally focus decisions on trade-offs between policy and investment options, rather than developing co-benefits.

A growing number of cities have been turning away from such dis-integrated 'silo'-based approaches. The GIZ-ICLEI study (2014) draws upon practices from cities worldwide that have recognized the crucial inter-linkages between sectors to achieve efficiencies in water, energy, waste and food – the typical focus areas of Nexus attention – but also to achieve more sustainable urban transportation, more inclusive housing and employment schemes, improved social amenities and more effective biodiversity conservation.

As has already been established in other publications (Hoff, 2011; Martin-Nagle *et al.*, 2011; FAO, 2014; ICLEI and UNEP, 2013), integration across jurisdictions and sectors creates opportunities for viable water–energy–food/waste/soil Nexus solutions.[2] However, the discovery of further Nexus opportunities in an urban context, with an expanded scope of productivity objectives, cannot be left to coincidence. It requires a defined and relatively specialized approach. Therefore, the Urban Nexus approach is outlined as an action-oriented framework to identify Urban Nexus opportunities and to develop projects for their implementation.

From integrated planning and systems thinking to the Urban Nexus design[3]

Today's pursuit of the Urban Nexus reflects more than two decades of planning innovation and project experiments in the area of 'integration'. These efforts mark a continuous evolution of the understanding of cities and urban regions as complex systems of systems – as agglomerations of political, market, infrastructure, resource, legal and institutional, ecological, community and cultural systems that are connected and connecting on a worldwide basis.

Critics of modernist urban planning in the 1950s and 1960s, such as Louis Mumford and Jane Jacobs, bemoaned the disconnection between the 'physical plant' of the city (i.e. its infrastructure and built environment) and the city's essential social function as a platform for community and economic self-empowerment. Critics of modernist international development practices lamented the modern state's often violent undermining of marginalized urban communities as they advanced economically and sought political power through their self-built settlements.[4] The standards, regularization and improvements of modernist 'urban renewal' often left the poor even more marginalized.

From a different angle, the first environmentalists of the 1950s and 1960s highlighted the ecological ills arising from the separation of functions and flows in the city. In the early 1970s UNESCO's Man and the Biosphere Programme introduced resource balancing studies as a way to track resource flows, conversion processes and waste across urban regions as complex systems. Interdisciplinary urbanists, such as the biochemist and cyberneticist Frederic Vester, introduced concepts of complexity, feedback and closed loop design into urban theory.[5] One outcome was the search for solutions beyond linear resources-in/waste-out designs of urban infrastructures and processes, identifying opportunities to further optimize resource productivity and reducing environmental ills by cascading and cycling resource use, mimicking ecological energy and nutrient cycles.

The need for an Urban Nexus approach, therefore, has long been recognized. During the 1990s the term 'integrated' became a leitmotif in the fields of resource management, transportation and business planning and urban planning.

The spread of market liberalization policies and associated global market integration provided a further necessary impetus for experiments in more integrated planning and management. During the period between 1980 and 2000, integrated resource management was established in regional energy, water and solid waste entities. Local and regional government bodies also instituted watershed-based planning in many parts of the world. In the corporate sector, integrated business planning became a new field of management endeavour during the same period.

Reflecting the ever-increasing scale, complexity, and resource constraints of cities in a globalizing world, it comes as no surprise that 'integration' became a major theme in urban management. Stakeholder engagement, trust building and coordination became new management themes (Harrison, 2006, p. 190). The approaches varied from country to country, but the underlying premise of 'joined up' governance became a global trend. New forms of statutory planning were established, such as Integrated Development Planning (IDP) in post-apartheid South Africa, a strategic planning process involving the local, district and metropolitan municipalities over a five-year period to plan and coordinate their investments in consultation with a wide range of local stakeholders (Harrison, 2006, p. 186).

The introduction of the concept of sustainable development in the late 1980s instigated further experiments. In the 1990s the Local Agenda 21 (LA21) process endorsed by the 1992 UN Conference on Environment and Development introduced another form of multi-stakeholder integrated planning, subsequently called urban sustainability planning. The approach fostered a basic commitment to interdisciplinary planning and urban project design with a view towards systemic solutions that simultaneously addressed economic, social and environmental concerns. A 2002 survey by the UN Department of Economic and Social Affairs and ICLEI-Local Governments for Sustainability documented LA21 planning activities in more than 9,600 communities in 113 countries (International Council for Local Environmental Initiatives, 2002).[6]

Public sector reorganization followed the integrated planning trend. In the late 1990s, for example, new 'environmental services' departments were formed out of once-separate water, waste water and solid waste services units. Planning and economic development, previously distinct silos, were sometimes joined into single departments. Inter-departmental committees and matrix management approaches became familiar local government mechanisms. Metropolitan planning organizations were formed to coordinate and 'integrate' policies and strategies across the municipal jurisdictions of a region.[7]

However, integrated planning and management has not proven sufficient in the pursuit of sustainable development outcomes. In the case of Local Agenda 21 (LA21), for instance, many municipalities failed to integrate their LA21 initiatives within the conventional functions and units of local governments, confining implementation to a set of one-off projects or general awareness raising rather than a process of systemic reform. In this fashion, thousands of municipalities have completed integrated planning processes and have officially sanctioned integrated development plans. Nonetheless, demonstration projects aside, the modernist legacy of siloed, uncoordinated city planning and development is still dominant in many cities of the world.

In conclusion, what was missing in the pursuit of integration as a policy and planning pursuit was a method for identifying and evaluating new organizational and technical solutions to achieve the objectives of integrated plans; that is, the optimized use of human, financial and natural resources. Whereas integrated planning may identify the areas of potential, the solution itself – the mix of measures that enable the achievement of planning objectives – requires the design of alternative market signals, business models, systems and schemes that would be acceptable to stakeholders and address their diverse interests while also achieving integrated planning goals. Providing such a framework, the Urban Nexus approach suggests a way to move beyond integrated planning to a design-oriented practicable approach.

What is the Urban Nexus?[8]

> The Urban NEXUS is an approach to the design of sustainable urban development solutions. The approach guides stakeholders to identify and pursue possible synergies between sectors, jurisdictions, and technical domains, so as to increase institutional performance, optimize resource management, and service quality.
>
> It counters traditional sectoral thinking, trade-offs, and divided responsibilities that often result in poorly coordinated investments, increased costs, and underutilized infrastructures and facilities. The ultimate goal of the Urban NEXUS approach is to accelerate access to services, and to increase service quality and the quality of life within our planetary boundaries.
>
> (GIZ and ICLEI, 2014, p. 6)

The Urban Nexus approach scopes the opportunities for integration at the different scales of the built environment and its infrastructures, across a region's supply chains and resource cycles, and across the siloes of policy and operations within local, regional, sub-national and national jurisdictions. For that purpose, an Urban Nexus solution integrates at least two or more systems, services, areas of policy or operation, jurisdictions or social behaviours, in order to achieve multiple urban policy objectives and deliver greater benefits with equal or less resources. Urban Nexus solutions typically involve a set of coordinated measures that span the areas of technology, policy, planning, finance, business models, institutional design and communications – amounting to a 'solution set'.

Although the process for identifying a prospect and designing a solution may be transferable, Urban Nexus solutions developed for one place may not be transferable to another. Therefore, the Urban Nexus approach is fundamentally a process of solution customization, and depends on valuable input from all relevant stakeholders.

The Urban Nexus Development Cycle provides a strategic design process for collaboratively translating integrated policy and planning objectives into feasible projects, technical solutions and operations.

Examples given in the GIZ-ICLEI study show that the most innovative cities in development and public management have been those that first established design processes, capacities and institutions to tailor implementation solutions to their planning objectives.

The following is a brief outline of each stage of the Urban Nexus Development Cycle:

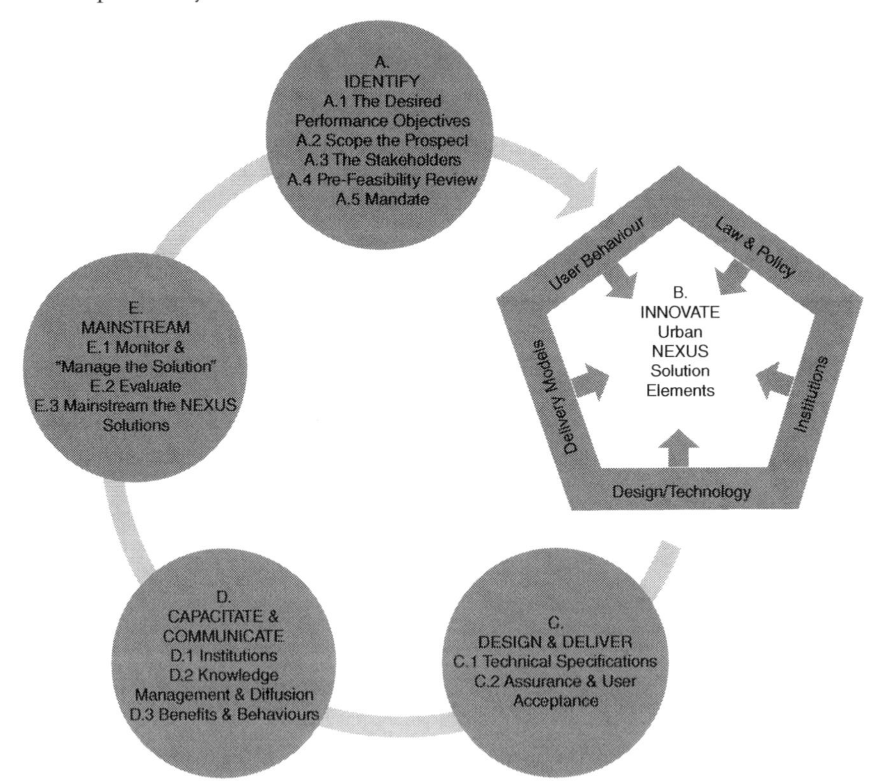

Figure 7.1 The Urban Nexus Development Cycle, a process for translating integrated planning objectives into policies, projects, systems and places. Source: GIZ and ICLEI (2014).

1. **IDENTIFY** First, adapt the four general Urban Nexus Practice Objectives to the local context – increased systemic effectiveness, increased demand-driven suitability and customization, increased productive efficiency, and increased resilience and adaptive capacity. Prospects for building synergies can be found by stakeholder scoping of five Areas of Integration:
   - Opportunities to achieve objectives through **Integration across Scales** of the built environment, infrastructures, local and regional supply chains and resource cycles, and policies and operations of local, regional, sub-national and national jurisdictions;
   - Opportunities to achieve objectives through **Integration of Systems** of resource extraction and power generation, food cultivation, processing, manufacture, resource supply and waste management etc. by establishing cascades and cycles of resources between systems;
   - Opportunities to achieve objectives through **Integration of Services and Facilities** to avoid the underutilization of valuable fixed assets by integrating services and facilities conventionally separated by sectoral functions;
   - Opportunities to achieve objectives through **Integration across Silos**, consolidating institutional interests and managerial and professional 'silos' arising from the organization of urban areas and systems into separate jurisdictions, utilities, and departments; and finally
   - Opportunities to achieve objectives through **Integration of Social Relations and Behaviours** to enable the engagement of all stakeholders in the above integration dimensions, and counter legacies of cultural, social and political division.
2. **INNOVATE** The identified stakeholders next collaborate in a structured innovation process to develop a set of politically, institutionally and economically viable measures in the identified and prioritized Areas of Urban Nexus Innovation. These measures typically span the range of law and policy, design and technology, delivery models and financing, communications and changing user behaviours, and institutional design and development.
3. **DESIGN and DELIVER** The design and delivery of the solution includes prototyping and piloting it in a real-world operating environment. Each individual measure is tracked and evaluated during the piloting to ensure that it is contributing as necessary to the combined performance of the integrated systems. This includes assessing and understanding the end user's response to the new solution.
4. **COMMUNICATE and CAPACITATE** Three main areas of capacity building are typically required to establish a new solution. These are:

- training of operational staff to manage their parts of the integrated solution;
- behavioural change and skills development for end users and other beneficiaries so that they can effectively use the new solution; and
- the establishment of a systematic process for introducing and supporting the solution in new locations or facilities, including through institution-building measures.

5 **UPSCALE and MAINSTREAM the Urban Nexus** Mainstreaming in many cases is a matter of designating or creating an entity that specializes in the scaling of all the unique aspects of the given Urban Nexus solution – an entity with the capacity to address location-specific problems and to 'manage the solution' within different contexts.

Lessons from application of the Urban Nexus framework

To illustrate the potential of an Urban Nexus approach, the GIZ-ICLEI report reviewed thirty-seven case examples of existing projects that apply the five Urban Nexus objectives to varying degrees at the scale of a facility, a city or an urban region. These examples cover the full range of instances in which municipalities and NGOs are using integration measures to address social and economic productivity and more optimized use of land, facilities and natural resources.

The cases range from sanitation, housing and energy and food supply to climate change mitigation and biodiversity protection. In eThekwini/Durban, South Africa, for instance, the Mariannhill Landfill Conservancy illustrates how Nexus design approaches can be applied to a solid waste landfill.

In the first instance, the landfill was redesigned as a closed-loop system to prevent toxic material from contaminating the surrounding area. The Mariannhill landfill also controls the release of greenhouse gases (i.e. converting methane into CO_2) through a gas-to-electricity plant. The power produced is fed into the local grid. Thanks to integrated design, the landfill now also serves as a public recreation space for the community and as a conservancy for indigenous plants (aided by a Plant Rescue Unit) and a registered national birdwatching site.

In addition, the development of a Mariannhill Landfill Conservancy involved a high level of public participation. A Monitoring Committee involves the community and environmental organizations in the maintenance of the Conservancy. Providing a new public space to the city, it also contributes to public awareness concerning waste management by offering educational workshops and school visits.[9]

In addition to the case studies, in 2014 GIZ and ICLEI applied the Urban Nexus principles to two pilot projects in Nashik, India and Dar es Salaam, Tanzania. These cities took the first steps in implementing an Urban Nexus approach in an exemplary way. Over the limited duration of the pilot projects the Urban Nexus brought together a wide range of stakeholders who had never before sat together at one table, thus generating a new 'institutional nexus'. They collaboratively designed and implemented innovative solutions and programmes for optimizing water, energy and land resources in peri-urban agricultural practices (Nashik), and improving the learning environment at two municipal schools while installing integrated energy-efficient technologies, rainwater catchment and vertical food production systems (Dar es Salaam) to demonstrate the benefits of Urban Nexus thinking to pupils, local communities and government officials.

Factors for success identified in these projects and case studies done for the GIZ-ICLEI report include:

- Identify 'hotspots'. When identifying priorities for Urban Nexus projects, consider the areas, or 'hotspots', where this approach would have the most multiplier or ripple effects to maximize the reach and benefit of the initiative.
- Bring all stakeholders around the same table by creating 'Urban Nexus Task Forces'. Urban Nexus Task Forces created to oversee Urban Nexus projects at the urban and regional level serve the purpose of linking relevant departments and levels of government together with other key stakeholders (experts, civil society, private business, NGOs and multi-lateral organizations). Urban Nexus Task Forces are a simple way to kick off, strengthen and sustain cross-departmental collaboration offering stakeholders a taste of 'breaking the silos'. Eventually, the goal is to institutionalize such multi-stakeholder collaboration.
- Encourage governmental authorities and stakeholders at all levels to be part of Urban Nexus solutions, which should re-connect scales and optimize complex cross-boundary resource flows (e.g. river basin management).
- Promote supportive framework conditions for Urban Nexus solutions at all levels. Regardless of their size and scope, Urban Nexus projects are embedded in regulatory and administrative frameworks. For example, national 'silos' in regulation, public procurement, budgeting and accounting processes, etc. can hinder innovative integrated approaches and cross-departmental cooperation at the local level. Supportive national and decentralized frameworks regarding legislative mandates, financial support and incentives are therefore crucial for the scaling up of successful local and regional Urban Nexus initiatives.

(GIZ-ICLEI, 2014, p. 9)

Conclusion: an opportunity to link global development goals and objectives with the world's urban realities

Over the last decade governments at all levels, international institutions, the private sector and NGOs have themselves developed a 'nexus' of common purpose regarding the latent potential of cities. This is reflected in the goals of the international community, in the evolving urban development policies and strategies of many nations, and in the advancing best practices of metropolitan regions.

Collectively, the above-mentioned actors have long recognized the adverse global consequences of uncoordinated, fragmented approaches to urbanization. One often-cited indicator is the persistence of a population of nearly one billion 'slum dwellers' in the world, proving the inadequacy of past urban housing, infrastructure and services approaches.[10] Inefficiencies in urban energy provision and demand-side management have made urban regions the point sources of global climate change, with as much as 80 percent of the world's greenhouse gas emissions generated in cities (World Bank, 2010). Even the positive benefits of urbanization, such as access to higher protein diets, are associated with the degradation of the world's fisheries and sensitive terrestrial ecological systems.[11]

Reflecting this reality, the international community has increasingly recognized the importance of cities and the roles of regional and local governments as partners in addressing global challenges. This recognition began with the United Nations endorsement of Local Agenda 21 in 1992, the Habitat Agenda in 1997, the establishment of the Cities Alliance in 1999, and with the engagement of local government associations in the international climate and biodiversity convention processes. Even as cities are places where global challenges are aggravated, they also hold great potential as places of productivity, creativity and efficiency.

However, urban governments and stakeholders have hardly developed their cities to their potential as centres of social, economic and environmental problem solving. We still systemically underutilize the full productivity potential of our cities. It is towards this end that the Urban Nexus approach is being developed by a wide range of institutions, from local governments, NGOs and international city networks to national governments and international organizations like the UN Food and Agriculture Organization (FAO).

Applications of the Urban Nexus approach in individual cities and metropolitan regions may start small, as stakeholders learn how to work together differently. However, at scale, Urban Nexus solutions can have profound impacts. These are illustrated by the case studies presented in the GIZ-ICLEI study (2014). Collectively, the cases selected for this study indicate how Urban Nexus solutions can not only address resource challenges in developed cities, but also contribute to the ultimate achievement of international development objectives, such as the United Nations' Millennium Development Goals and the 2015 Sustainable Development Goals.

Notes

1 GIZ and ICLEI-Local Governments for Sustainability (2014) *Operationalizing the Urban NEXUS: Towards resource-efficient and integrated cities and metropolitan regions*, by Jeb Brugmann, Kathrine Brekke, and Lucy Price (Eschborn: GIZ/ICLEI, 2014). Available online for download at www.iclei.org/urbannexus

2 The global survey found that integrated and participatory governance and innovation are the most important success factors for projects/programmes for resource efficiency in cities.

3 This section is based on and extensively draws text from the Brugmann and Flatt (2014) background paper for the GIZ-ICLEI study (2014).

4 See for instance Turner, J.F.C. and Fichter, R (1972) *Freedom to Build: Dweller Control of the Housing Process*. Macmillan, London. Other origins of the critique of modernist development practice *by development practitioners* can be found in the writings of Mary L. Elmendorf, John Habraken and Arif Hassan.

5 See Vester, F. (1977) *Ballungsgebiete in der Krise* (Urban Systems in Crisis). DVA, Stuttgart.

6 In the 1990s and early 2000s GIZ played a fundamental role in the establishment and mainstreaming of Local Agenda 21 practice in Asia and Latin America in particular, establishing national training and demonstration programmes and working closely with and providing financial support to ICLEI.

7 However, the formation of consolidated inter-municipal bodies for multi-level governance of metropolitan regions generally failed apart from in top-down central government-controlled contexts such as Brazil or China, where municipal limits were simply expanded. The European compromise was sought with other mechanisms for inter-municipal cooperation.

8 This section is based on and extensively draws text from GIZ and ICLEI, 2014, pp. 6–9 by Brugmann, Brekke and Price.

9 Find the full Urban Nexus Case Study No. 6 on eThekwini/Durban and many other cities at www.iclei.org/urbannexus.

10 The term 'slum' usually has derogatory connotations and can suggest that a settlement needs replacement, or it may legitimate the eviction of its residents. It is a difficult term to avoid, but it is not used here with derogatory connotations. Where the term is employed in this study, as per the definition used by the International Institute for Environment and Development (IIED) it refers to settlements with some of the following features: a lack of formal recognition by local government of the settlement and its residents; the absence of secure tenure for residents; inadequacies in provision for infrastructure and services; overcrowded and sub-standard dwellings; and location on land less than suitable for occupation. For a discussion of more precise ways to classify the range of housing sub-markets through which those with limited incomes buy, rent or build accommodation, see *Environment and Urbanization*, 1(2), available online at: http://eau.sagepub.com/content/1/2

11 Other ecological stresses are tied to wider trends of population, consumption, and industrial inefficiency. These include water stress, peak soil and species losses. For instance, water stress is caused by the unequal distribution of water on the planet, as well as by a rate of water use that has grown twice as fast as the rate of population increase over the last century.

References

Brugmann, J. and Flatt, J. (2014) 'From integrated planning and systems thinking to Urban NEXUS design', *Urban NEXUS Background Paper*, ICLEI and GIZ, Germany.

FAO (2014) *The Water-Energy-Food Nexus: A new approach in support of food security and sustainable agriculture*, Food and Agriculture Organization of the United Nations, Rome.

GIZ and ICLEI (2014) *Operationalizing the Urban NEXUS: Towards resource efficient and integrated cities and metropolitan regions*, GIZ Eschborn, Germany. Available online at: www.iclei.org/urbannexus

Harrison, P. (2006) 'Integrated Development Plans and Third Way Politics', in U. Pillay, R. Tomlinson and J. Du Toit (Eds.), *Democracy and Delivery: Urban Policy in South Africa.* Cape Town: HSRC.

Hoff, H. (2011) *Understanding the Nexus.* Background Paper. Available online at: http://www.water-energy-food.org/en/news/view__255/understanding-the-nexus

ICLEI and UNEP (2013) *Towards Resource Efficient Cities: A global city survey.* Available online at: www.iclei.org/fileadmin/PUBLICATIONS/Papers/GI- REC_Global_Survey_Report_Final_2013_ICLEI-UNEP.pdf

International Council for Local Environmental Initiatives (2002) *Second Local Agenda 21 Survey, Background Paper 15, Commission on Sustainable Development, World Summit on Sustainable Development, Second preparatory session.* UN Department of Economic and Social Affairs, New York. Available online at: www.un.org/jsummit/html/documents/backgrounddocs/icleisurvey2.pdf

Martin-Nagle, R. *et al.* (2011) 'The Water, Energy and Food Security Nexus – Solutions for the Green Economy', Bonn 2011 NEXUS Conference Synopsis. Available online at: www.water-energy-food.org/en/whats_the_nexus/messages_policy_recommendations.html

Turner, J.F.C. and Fichter, R. (1972) *Freedom to Build: Dweller control of the housing process.* Macmillan, London.

Vester, F. (1977) *Ballungsgebiete in der Krise* (Urban Systems in Crisis). DVA, Stuttgart.

World Bank (2010) *Cities and Climate Change: An urgent agenda*, The World Bank, Washington DC, p. 15. Available online at: http://wbi.worldbank. org/wbi/document/cities-and-climate-change-urgent-agenda

8 The confederacy of experts

The crushing Nexus of silos, systems, arrogance and irrational certainty

Gary Lawrence

Introduction

Today we recognize that solving the Nexus of Water, Energy and Food is critical to the successful survival of *Homo sapiens* in the coming decades. The critical resources that sustain our agricultural production – water, energy and land – are increasingly under threat from population growth, climate change and habitat loss or contamination. Unfortunately, our endemic tendency to distance ourselves from overly complex issues means that we are overlooking the most important factor in the debate, the very factor that makes this problem so critical: people. We must remember why it is important to solve the Water Food Energy Nexus. It is not because the planet is at risk, but rather that the ability for nine billion people to live securely and harmoniously on the planet depends entirely upon these essential ingredients for life.

The problems of our time, belief in the value of specialization over generalization and the history of educational praxis inspire people to become ever more "expert" in their chosen field – to delve deeper into technical understanding, to focus more narrowly on one specific application, to create niches within already specialized niches. Our expertise in the fields of water, energy and agriculture is growing more sophisticated every day. We can analyze, predict, model and innovate at an almost molecular level in each field. What we are not good at is the big picture.

When identifying issues or providing advice, brilliant, well-intentioned individuals typically define problems within the scope of their own understanding and cognitive biases. Seldom do they consider broader contextual issues. Even more rarely do they engage the wisdom of the masses, the stakeholders who are most directly impacted by the issues. We succeed only in fixing an attribute of the problem while the root cause festers and propagates beneath an elegant technical solution.

Our world is urbanizing, changing far more rapidly than our experience and understanding are able to accommodate. People are torn between a desire for nostalgic continuity and the opportunity to take beneficial advantage of changes that seem benign or advantageous. Given that we know things are changing, but not how fast or to what extremes, today's most responsible answer to most things is "*we don't know, we'll have to figure it out.*" A lack of constructive humility, our lack of ability to admit that we don't know but we must learn, may be society's most dangerous flaw.

The tyranny of experts

Over the years Nero and Tiberius have given way to a global host of dictators and oppressors whose deeds fill our media channels on a daily basis. These are easy targets to single out. The plight of their people, the destruction of towns and cities, the exploitation of natural resources to benefit the few while the majority goes hungry and homeless, all point back to the tyranny of decisions made by one man, one woman or one elite group. Sitting in the democratic West with our freedom of speech it is easy to condemn, to moralize and to point out the right and wrong of the situation. However, it is not so easy to recognize a more ubiquitous tyranny in our own midst.

As noted earlier, our culture and our education systems compel people to become ever more "expert" in their chosen field; to delve deeper and deeper into technical understanding and focus increasingly narrowly on one specific area of application, creating specialist niches within already specialized fields. My own company, AECOM, abounds with such specialists. We are proud of them and our clients recognize their value. Technical excellence is the essence of our culture. However, we no more believe that one type of technical expertise should decide the fate of an entire city's infrastructure than we believe a single individual should decide the fate of a country.

Identifying the fundamental problem

The most spectacular failures in my career have been caused by racing off to solve a problem I identified as critical, only to find that it was no one else's priority or no one else had any sense of urgency about fixing it. It is too easy for professionals to propose solutions to problems they think need solving without pulling their heads up for long enough to look around and ask the people most affected by the problem what they think.

I came across one example recently that illustrates this point beautifully. An entrepreneurial team of students with a desire to make a difference in the world came up with a clever design for a soccer ball that kinetically generates power while it is being kicked around. At the end of the day young students in impoverished communities can pick up their ball and plug it into a light so that they can see to do their homework. This is a wonderful idea and indeed fills a need. However, when the reporter I was listening to questioned adult members of the South American community where the balls had been distributed, she learned that the power grid did in fact extend to the village, it was just that no one had any money or resources to tie the houses into it. Without in any way wishing to diminish the vision and compassion of the inventors, one has to wonder if the money spent in development and distribution would not have been more effectively and sustainably spent on extending the existing grid into the villagers' homes. The root cause of the problem was not technological, it was political and financial.

These types of situations often arise as the result of not asking the right question. They are facilitated by the tyranny of experts. Those people so focused on a technical innovation, a personal goal or a conventional, hitherto successful approach that they fail to appreciate and incorporate the wisdom of the masses. If we do not listen to a broader constituency of stakeholders who are able to provide unbiased information accessible uniquely to them, we are necessarily going to miss the opportunity to solve a persistent problem, rather than simply fixing an attribute of it.

Around the world there is an increasing number of high-value fixes being proposed for the problems of climate change and urban growth. Building physical resilience in cities often focuses on massive engineering projects, expensive retrofits and additional infrastructure to supply water, energy and food. In almost every case, any proposed solution meets with strong public resistance. This is inevitably because investments can only be made in one area of concern at the expense of another.

For every investment there is a high opportunity cost. Every stakeholder is juggling multiple priorities, and they must decide between reducing taxes, fixing more pressing problems or preserving their political capital. The public has its own ideas about what is required and what its money should be spent on. If we cannot effectively articulate how the investment can improve lives, create value and reduce risk, our solution, no matter how brilliant, is just a castle in the air.

A question of capital

Solving such problems requires a level of integrated systems thinking that is simply not possible in our current economic market circumstances. The economic system, an artificially created construct that we invented initially as a way to reduce conflict over wealth, has evolved into a way for the winners to aggregate most of the wealth at the expense of those who will never have the resources to compete. It is this unwavering focus on financial capital that has blinkered us to the importance of the three other types of capital upon which success is founded.

Social capital is the result of creating healthy, equitable civil environments that engage every citizen in societal life and offers each individual the opportunity to be successful on his or her own terms. It is only by creating such environments that we allow the possibility to build human relationships, networks and the trust that allows business to succeed.

Natural capital encompasses all Earth's natural resources and the ecosystems that sustain life: air, water, earth, and energy in the form of the sun. There is not a business operating on the planet that does not rely on taking these natural resources and converting them to some other purpose in order to generate profit. There is not a life on this planet that does not depend upon clean air, clean water, uncontaminated earth and clean energy for survival. Earth will continue with or without us – our lives and livelihoods depend upon responsible stewardship of our finite natural capital.

In stark contrast, the only natural resource that actually increases over time is human capital. The limited way to think about this is in terms of population growth and labour. The more critical aspect is the expanded intellectual capacity that humans can apply to addressing the complexities of life now and in the future. When the human mind is given the space to be curious and the opportunity to explore, great things become possible. This does not happen when the mind is held confined within its designated silo, fixated upon a specific technical solution, financial mechanism or desired social outcome.

Reaching agreement

When professional, public and governmental silos come together to act, each approaching a problem from its own area of expertise, the resulting action is often based upon a compromise. This usually means that the interested parties have agreed upon how to address some of the root problem's attributes, but rarely does it mean that everyone shares a common understanding of the fundamental causes of the problem. Solving today's complex problems, in the hope of creating the conditions in which we can optimize human development over time, is fundamentally problematic.

The best example of this comes from a recent meeting of the United Nations Economic and Social Council. The discussion revolved around the question of how the government and private sectors could work together to increase disaster resilience and eradicate poverty. The disconnect here is that the public sector believes the private sector has a moral obligation to solve these sorts of problems. Nothing can ever be achieved from this moral viewpoint. Business is in business to benefit business. Business also generates most of the wealth that needs to be applied to solving the ever-evolving complexity of human systems. In reality, without the wealth being generated by the private sector, the meeting that brought all of us together that day to ponder such problems would not even have been possible. Private sector taxation is the source of government funding. Clinging to the perception that the private sector is not a legitimate player in optimizing conditions for the future and mistrusting private capital will prevent the government sector from making any progress at all.

The question is not one of moral obligation, then, but rather one of self-interest. Government can engage in meaningful conversation with the private sector if instead it asks, "What are the benefits for the private sector in helping society be more successful?" Then, rather than moral obligation, the question becomes one of degree. Is the private sector keeping too much of the wealth deriving from the use of natural resources for which it paid nothing? Does the health burden of contaminated resources outweigh the cost of responsible resource management? What are the financial consequences of resource depletion or unpredictable availability to business operations? Are we investing enough in educating individuals to develop the technical and entrepreneurial capacity necessary for the private sector to be successful into the future?

None of these questions are "either or." They are all "and both" questions, matters of degree. It is firmly within the interests of the private sector to ensure that the resources and human potential necessary to keep the private sector functioning are protected and improved. The private sector does not need to be morally outraged to care about the poor, about education, about resource degradation or about the impacts of climate variability. For the private sector these issues are well understood within the context of risk assessment.

Knowing what to do

Design can go a long way towards addressing aspects of some problems, but it cannot address the fundamentals. The fundamentals are political, and we must be honest with the public about how confident we are in the solution and what its limitations may be. Based on conventional wisdom and business as usual there is inevitably greater confidence in addressing attributes of problems, but these are essentially short-term fixes and patches – if we recognize them as such it is easier to have an honest conversation about trying to effect more meaningful, long-term improvements.

I will illustrate this point from my observations at the Nature Conservancy's Learning Exchange on the topic of "*Mainstreaming Climate Resilience: How do we get there?*" Two types of people were present: the green infrastructure people, and the grey infrastructure people.

The Green People believe that constructing infrastructure from concrete and other aggregate materials creates a hardscape that essentially exacerbates climate problems because it diverts and waylays the problem rather than letting the properties of the natural environment deal with it where it arises. In the aftermath of Superstorm Sandy the Green People stepped forward to argue that we should not attempt do anything with dykes and draining, but rather leave the beaches alone to replenish themselves over the generations to come.

The Grey People believe that hardscape construction is the only viable answer and that human intellectual and engineering capacity is stronger and more predictable than letting nature take its course. They believe that the only viable response to potential future storms is to design and construct storm surge barriers.

In the minds of both the Green and the Grey there was no room for doubt. The conversation became an ideological struggle between two sets of experts, each striving to convince the other that their way was right.

Yet our only possibility of success begins with agreeing that both Green and Grey are committed to the same cause. The distinctions between Green and Grey are distractions. What are manmade materials if not the result of a conversion of natural resources? What industry exists without an abundant source of clean water and an equally abundant source of energy? How can human ingenuity thrive without the clean air, water, food and energy essential to human survival? No species can exceed its range. Surely we can all agree on that?

There is no perfect answer. There is a best answer, which is economically viable, technically feasible and politically acceptable, but this is only possible if the Green and the Grey are willing to come together, to listen to each other's ideas, and to find integrative rather than opposing solutions. Holding out for the perfect answer is an avoidance tactic that leaves a lot of people in harm's way.

It is easy to point to the professions and conclude that they are filled with stubborn experts, but in reality they are only part of the problem. Every neighbourhood boasts its own "experts." In every neighbourhood there is a vocal minority capable of mobilizing resistance to any threat to the status quo. Invariably the status quo in question is a sanitized vision of a perfect past that may or may not have existed.

In my days as Planning Director for the City of Seattle I met many of these people. They would emphasize the utopia that their neighbourhoods used to be before higher population density, more cars, more restaurants or cinemas. They talked about an almost bucolic existence of freshly baked bread and friendly neighbours – a vision that was a very far cry from the city's past reality as a pioneer town rife with fleas, polio, typhoid fever and brothels.

Another example from my planning days shows just how the conversation changes when you engage a broader audience. We were trying to figure out why bus ridership among women declined during the winter months. The transit experts (nearly all male) decided that it was because there were not enough bus shelters and women don't like getting wet while waiting for the bus. Fortunately we did not take their word for it. We engaged some psychologists from the university to examine the problem. What we learned was that we were not dealing with an issue of physical discomfort, but rather with an issue of fear. The real problem was related to lighting. Women were afraid to walk to and from the bus stop in the dark mornings and evenings because there was inadequate street lighting. This revealed a completely different course of action, and one that was much more cost effective.

There is always a balance that can be found. It is never perfect, but it acknowledges that everyone's perspective is valid and rational from his or her point of view. If we can begin with what we all agree upon – clean air, water, food and energy – we will recognize that we are all striving towards the same goal. We are all people with hearts and souls and the capacity to honour and celebrate the resources necessary for life.

Choosing to act differently

Urbanization, population growth, climate uncertainty, technological innovation and resource limitation are accelerating at an unprecedented pace, throwing more and more uncertainties and unknowns into the equation. Yet too often we are certain we know the answer without having followed any sort of methodological approach to gaining knowledge. We rely heavily on past experience and conventional wisdom when it is nothing short of irresponsible to assume that we know the answers to such complex problems.

The people most invested in the status quo are the self-proclaimed "experts." When the status quo is challenged these people can no longer be considered experts. This is the greatest barrier to rational action on the big issues of our time. When their expertise is challenged the experts fight to control their domain. The public is expected to simply defer to the intellectual superiority of others who claim the high ground.

We have to admit that we don't have the answers. The best answer possible today may be very different from the best answer possible tomorrow. The only attitude that makes sense here is one of doubt. Nevertheless, we still have to make progress. Humility is the quality that will allow us to start down a path but recognize if it's a mistake, and find the courage to change paths.

We must find ways to communicate what is really at stake in such key decisions. We must move beyond conventional wisdom into actual rational conversations about cause and effect. The final requirement is that those who are empowered to act choose to do so. The decision to act or not is a question of political, financial and resource risk management. But it is also invariably fraught with the more human influences of culture, nostalgia, aspiration, fear and what Francis Bacon described as the "*preference for truths that we would rather believe.*" Good design and good science can help reduce the risk. Good outreach and communication can pave the way to making such decisions more possible.

Intelligence is not the bastion of an academic elite. It is evenly distributed among the population, not held in an expert class. Everyone is an expert in something, but no one is an expert in the future. We need expertise, but we also need the "hive-mind" – the collaborative processes that lead to better understanding and problem solving. We can anticipate scenarios, "nudge" in better directions, adapt to urgent realities, and we can much more rapidly discard errors in judgement, but we cannot know the future until it happens. The hive mind is a powerful tool – and one we ignore at our peril.

Security: The overlooked Nexus issue

Even the phrase "Food Water Energy Nexus" fails to capture the ultimate significance of our dilemma. Water and energy are integral to the cultivation, production and distribution of food, but food is fundamental to our security.

In eighteenth-century Paris the first thing the Lieutenant of Police did every morning upon reaching his desk was to inquire as to the price of a loaf of bread in the city's bakers' shops. The price of the loaf would tell him whether he could expect a quiet day, or whether he should prepare for a day of unrest and violence in the streets. By the time the Bastille was stormed on July 14th, 1789 a loaf of bread cost the average labourer an entire month's wages. From the French Revolution via the Russian Revolution to the American Revolution, the cost of food as a catalyst for change cannot be ignored. Indeed, the International Food Policy Research Institute believes the destruction of

Russia's wheat harvest in 2010 and the subsequent food price spikes in Egypt helped to trigger the events now known as the Arab Spring.

In a report published in 2013 by the Food and Agriculture Organization of the United Nations entitled "The State of Food Insecurity in the World", the overriding conclusion is that international price volatility and high food prices are expected to continue in the years ahead. Food production is largely understood as agriculture – which requires land, energy, water, and chemicals for fertilizer. Food availability, on the other hand, encompasses ports and shipping, rail systems, trucking, refrigeration, warehousing, storage, supermarkets, packaging (more chemicals), transporting consumers to and from markets, home refrigeration and, particularly in developed countries, a huge amount of waste disposal.

The movement of food from farm to market to shopper is a major consumer of energy, but those energy costs are not captured in the price of the food on our shelves. As we run out of arable land this will change, prices will rise and cities must become part of the solution.

Humility: The way forward

As we think about our future, and what we need to do as governments, corporations, planners, designers, builders and citizens to protect and sustain our communities, we must recognize two things:

1. We are, in fact, all in the agriculture business
2. Not one of us is an expert in knowing the future.

In evolutionary terms our real purpose is to survive and enhance our existence. Our current behaviours are threatening that purpose. We are creating conditions in which we may not survive, but a belief that technology will save us and deferring to "experts" to solve the problem is a vanity we cannot afford. Our challenge is to be humble, to fix what we can, and to adapt our behaviour in full recognition that nature has a store cupboard full of surprises. Humanity is not going to win a fight with nature, but humility may allow us to coexist.

Part 3

Natural resource security for people

Water, food and energy

9 Water–food–energy–climate

Strengthening the weak links in the Nexus

Ania Grobicki

Introduction

The geopolitical–economic–social underpinnings of the global systems of food, energy and water are increasingly interconnected. While the concepts of food security, energy security and water security have been defined, they are still often framed as being linked to resource scarcity, human security and poverty eradication, neglecting the major climate risks we now face. Advocates continue to put forward rather static technical solutions and policy frameworks to secure water, food and energy for growing populations, based upon received knowledge and best practices.

In the age now termed the Anthropocene (Zalasiewicz *et al.*, 2008), we may well believe that these global systems are under human control. However, there is a bigger picture which introduces elements of greater risk and uncertainty into the "anthropogenic" Nexus of food–energy–water. Water-related disasters such as floods and droughts bring sharp reminders that these interlinked global systems are highly vulnerable to climate change. Global environmental sustainability is a far distant goal. War and disease are still ravaging the world. Some nations and societies are failing, falling outside the reach of global systems.

The water and climate crisis we are facing requires a deeper analysis, which includes the "natural" Nexus. We need to be clear about the limits to human management of global systems, and find the weakest links. This may require a reframing of the issues away from the current economic focus on resource scarcity, towards adaptive societies and systems that help build sustainability through risk reduction and resilience. This chapter seeks to explore these issues.

Three major resource systems that are socially defined

Food – from subsistence farms to smallholder agriculture to large agribusiness, producing food is largely a private undertaking. The massive system of global food trade is partially regulated by public policy, international trade agreements, food safety rules and tariff barriers, but the trade itself is largely in the private sector, controlled by a handful of very large companies. The world food system is essentially anthropogenic. The production of food can fluctuate but is largely under human control, governed by social choices, and hence will continue to increase.

Human food security is very intricately interconnected with the major resource systems of energy and water. Currently 40 percent of agricultural production is irrigated, while 60 percent is rain-fed (Faurès *et al.*, 2007). Food production is boosted by the required quantities of water provided at the right time for optimum crop growth. The need for more food does not necessarily mean increased quantities of water, but rather a more precisely-targeted delivery of water to crops. However, increasing productivity in the food system does require increasing inputs of energy for mechanized farming, for producing other elements of the supply chain such as fertilizer, and for enabling the global food trade and distribution system.

World food and feed production needs are projected to increase by 60 percent by 2050 (FAO, 2012), although the world population is only set to grow by some 40 percent at most. The increase is due largely to the projected changes in diet to include more meat – a social choice for many people as they become more affluent. In turn, this means more water, energy and feed requirements for greater numbers of livestock. While land as a resource is limited, there are large productivity gains to be made in food and feed production and reducing food waste, which could mitigate much of the additional resource requirements for water and energy.

Energy – from small-scale renewables to large hydropower and large-scale fossil fuel use, there is a mix of private and public sector investment. Continent-wide smart grids are affected by government regulation and public policy, but the investments are also largely private. Essentially the energy system is an anthropogenic system. The actual quantities of energy produced are under human control, but with unlimited amounts of "renewable" energy from natural sources potentially available to be harnessed. Many forms of energy production require water within their processes, e.g. hydropower, thermal power production and fracking. Here a distinction needs to be made between those that merely use the water on a once-through pass (such as hydropower), later releasing the water downstream for other uses, and those (such as fracking) which consume water or pollute it, making it unavailable for other uses.

World energy needs are also rising fast, if not as fast as food needs, since technologies are becoming increasingly energy efficient. People are very resistant to energy price rises, yet higher prices bring about technological innovation and energy-lean devices. At present the increase in world energy needs to 2050 is projected to rise in line with the world's population, by some 40 percent.

Water – the management of water resources, from community-level supply and large urban water utilities to river basin management and national water policy, is largely a public sector undertaking with some private sector involvement. This makes the water system markedly different in its governance requirements to the food and energy systems.

There is also another essential fact to note, which makes water as a resource different to both food and energy. The total amount of water on Earth is essentially constant and therefore limited, if we include all forms of water (solid, liquid and gas) in all its locations, whether underground, in the soil, on the surface or in the atmosphere. Water cannot be "created" or produced. The natural water cycle determines the distribution of water between ice, freshwater and seawater (not forgetting atmospheric water and groundwater), through evaporation, precipitation, freezing and melting processes. However, ever larger parts of the natural water cycle are coming under human control, and these are the parts which are regarded as a resource.

Currently agriculture consumes between 60 percent and 80 percent of available freshwater in most countries, while energy needs, industry and household uses divide up the remainder – often leaving little for aquatic ecosystems such as rivers and lakes. Many rivers no longer reach the sea, while inland waters such as the Aral Sea are drying up or have already disappeared. Increasing the productivity of water in agriculture, producing "more crop per drop", is the focus of a great deal of research, and substantial productivity gains are being made in this area through technologies such as drip irrigation, remote sensing analysis that enables accurate targeting of fertilizer and moisture needs, etc.

The provision of water to human needs and activities is becoming increasingly commoditized. For instance, water supply is no longer solely a function of the quantity of surface water or groundwater that is naturally available, but additional supplies of "renewable", usable water can be created both through wastewater treatment technologies and through desalination of seawater and brackish water. However, these are still relatively minor contributions to water availability, except in water scarce regions.

Seawater desalination is an expensive and energy-hungry option for providing more water, although the technology is rapidly becoming cheaper and more energy efficient, a boon for drinking water supply to coastal populations. Saudi Arabia already supplies 80 percent of its drinking water from desalinated seawater. However, this is not a feasible option for agricultural water uses due to its cost. Some farmers in coastal regions are experimenting with salt-tolerant crops in order to get around the need for freshwater and salt-free soils. On a much larger scale, there are many farmers adjacent to large cities who are irrigating their crops with urban wastewater as an input, utilizing both the water and the nutrients that it brings. While testing and treatment should be undertaken to ensure that the wastewater is free of pathogens and toxins, the wastewater is an important asset to these farmers, for which they are prepared to pay (Bahri *et al.*, 2010).

The idea of water reuse for drinking water supply is also rapidly gaining more social acceptance. Cities such as Singapore and Windhoek in Namibia are leading the way, with the bulk of their water supply coming from highly treated "renewable" water. Its contribution will inevitably grow as more communities and industries realize that they can hold on to the water that they already have, and re-use it multiple times.

Food production should therefore be regarded as an end user of treated water, which has already been used as much as possible. We are seeing the development of an "anthropogenic" water cycle, which coexists with and augments the natural water cycle. While some project a water resource gap of 40 percent by 2030, this is unlikely, since water consumption can be delinked from economic growth as has already happened in Europe. Increasing water efficiency and multiple reuse will broadly keep water needs in line with population growth (less than 40 percent by 2050) or even below.

The "anthropogenic" Nexus

As we have seen, food and energy systems are both anthropogenic systems which are largely in private hands, while water when seen as a resource is more mixed (both public and private) but also largely under human and social control. In the "anthropogenic" resource Nexus, food, energy and water resources are closely interlinked through human design and action.

The consequence of the policy and investment choices that societies are now making can be dramatic, since the linkages in this Food–Energy–Water Nexus can be synergistic (mutually beneficial) or dysnergistic (mutually destructive). As examples, single-purpose dams are often dysnergistic, while multi-purpose dams are synergistic. Similarly, first-generation biofuels are dysnergistic while second-generation biofuels are shown to be environmentally sustainable.

Multi-purpose dams give a very clear picture of the anthropogenic Nexus in action. When they are designed and built for a single use, such as hydropower or irrigation, dams can frequently be seen as more damaging than beneficial. Over 70 percent of the world's dams were built for a single purpose, with 49 percent of these built solely for irrigation (ICOLD, 2014). Of the world's twenty tallest dams, nine were built solely for hydropower generation. However, in recent years many more multi-purpose dams have been built, providing benefits including water storage and drinking water supply, hydropower, irrigation, flood control, navigation and recreation. The operation of such multi-purpose dams can be difficult and challenging, sometimes with conflicting objectives – e.g. more water releases for hydropower are needed in winter, to generate electricity for heating, while more water releases need to be made in summer for irrigation and for ecosystem survival, when crops are parched and downstream wetlands and deltas dry up.

The challenge is to find synergies and to devise an optimum water release schedule that satisfies the various purposes for both hydropower and irrigation, which will take into account environmental needs downstream, e.g. fish spawning seasons and managed floods. An integrated approach is needed that takes all the potential costs, benefits and options into account. Overall, multi-purpose dams have been put forward as a compelling case for increased investment in dam building for development in many countries, overriding environmental concerns. Multi-purpose dams are now firmly part of the Food–Energy–Water Nexus, as defined by economic and social needs. Thorough policy dialogue and debate is needed in each situation, to find the optimum integrated solution for maximum synergy and multiple benefits.

Biofuels also illustrate many of the negative consequences of the Food–Energy–Water Nexus. Currently 2.7 percent of road transport worldwide runs on biofuels. Heavily subsidized and promoted as an environmentally sustainable route to energy security, biofuels contribute to the decarbonization of transport as the carbon they release is taken up during the crops' growing season – essentially a closed carbon cycle. However, crops grown to be processed into biofuels such as ethanol or biodiesel have been criticized for taking up scarce water and energy resources. The biofuels market has diverted land and crops away from food production, contributing towards driving up food prices as happened during the food price spike of 2008. When full life-cycle costs including fertilizer are included, the net energy benefit of some biofuels can be zero or negative. However, "second-generation" biofuels derived from crop residues have been shown to be much more sustainable and do not displace food production. Again, it is important to see biofuels within the "anthropogenic" Food–Energy–Water Nexus, and find the optimum integrated solutions for sustainability through social and policy dialogue.

What happens when we introduce climate into the Food–Energy–Water Nexus?

Introducing climate brings a substantial element of risk and uncertainty into the Food–Energy–Water Nexus. The IPCC is predicting that extreme climate events such as floods and droughts will continue to grow more frequent and more severe (IPCC, 2011). As long as the Nexus discussion is framed purely in resource terms, many of these systems are under human control. However, each of these major global systems is intimately interlinked with our climate, involving feedback loops that require natural systems to be brought into the analysis.

Food and climate – every farmer knows that the crops they grow are heavily dependent upon climate and rainfall. Floods and droughts can wreak havoc on food production and storage, as well as on food prices. The IPCC has found that temperate and cool regions will overall benefit from climate change, while food production in tropical and arid regions will drop, sometimes drastically (IPCC, 2011). Many subsistence farmers and the poorest of the poor in developing rural areas are the most disadvantaged by climate change. Currently agriculture and related land use changes are responsible for 24 percent of greenhouse gas emissions worldwide. Agriculture is a highly managed sector with extensive opportunities for both mitigation and adaptation.

Energy and climate – the direct link between fossil fuel use and climate change has long been established. Currently humans are removing hydrocarbons from beneath Earth's surface, burning them as fuel and hence putting carbon into the atmosphere at an increasing rate. In 1750, at the start of the Industrial Revolution, carbon dioxide levels in the atmosphere were 280 parts per million (ppm). In April 2014 they rose to 400 ppm for the first time, dismaying climate campaigners advocating decarbonization policies to keep the levels at 350 ppm.

Decarbonizing the energy system is now a top priority for human survival, but this is easier said than done when fossil fuels are so deeply embedded in our societies and economies. In 2010 the energy sector was responsible for some 35 percent of greenhouse gas emissions.

Water and climate – over and above the fraction of water that is socially determined as a human resource, water is a medium that pervades all life and the entire biosphere. The global water cycle is a natural part of the functioning of the planet's biosphere, increasingly affected both by anthropogenic activities and by climate change. For instance, floods are becoming more frequent with climate change. However, floods are also compounded by the anthropogenic changes to land use in catchment areas, paving over surfaces and draining wetland areas. These changes increase the proportion of rainwater which runs off over the surface of the catchment, causing floods downstream, instead of keeping water upstream and allowing it to infiltrate the soil. Unsurprisingly, the "hardening" of catchments results in much faster discharge of water downstream, with traditional flood defenses becoming overwhelmed.

Global climate change affects the global water cycle dramatically, as there is an increase of 6–7 percent in atmospheric moisture for each one degree of global warming (Wang, 2014). Speeding up the global water cycle in this way means faster evaporation, larger water losses through evaporation, greater risk of drought – and the converse, since what goes up must come down and the water vapour loading the atmosphere results in clouds forming more quickly and easily, rain falling harder and more erratically, more severe storms and greater risk of floods. "Hundred-year floods" are now becoming twenty-year floods, while twenty-year floods are becoming two-year floods in many parts of the world (Vorosmarty, 2014).

The past is no longer a reliable guide to the water cycle's future behaviour. In the language of hydrology, "stationarity" is dead. A "managed system" in terms of the surface hydrology of the earth, attempting to manage water resources through the design of built infrastructure including weirs, dams, stormwater drains and flood defences, has become partly unmanageable – not only because many of these anthropogenic defences have damaged ecosystems and natural water storage, but also because the risks of extreme events have increased and, more importantly, become highly unpredictable. Nature has entered the anthropogenic Nexus.

The "natural" Nexus

In the natural Nexus, food, energy, water and climate are interlinked through the manifestations of the forces of nature, such as hurricanes, floods and droughts. Drought is a classic Nexus issue – it is when drought develops that the interconnections and necessary trade-offs between water supply provision, energy production (especially hydropower) and food production become painfully apparent. Farmers are often the first to feel the economic effects of drought when their crops shrivel, and then when the drought persists deeper economic impacts spread through the Nexus. As rivers run dry, hydropower stations fall idle, reducing the power availability across the electricity grid. This can cause power shortages and blackouts which affect industries, hospitals, transport networks – and ultimately impact the whole state or national economy, as is frequently the case in India, and as occurred in South Africa in the drought of 2010.

At the other climate extreme, when flooding hits, there are also serious implications for farmers in terms of crop and livestock losses, as well as for cities, for infrastructure, including power stations and grids, for industry and for entire economies. The floods in Thailand in 2011 interrupted global supply chains ranging from computer hard-drives and motor vehicles to rice production, causing US$60 billion worth of economic losses.

Increased storm surges and sea level rise are putting at risk housing, livelihoods, industry, food production and energy production in the crowded, highly productive coastal areas and deltas around the world. In developing countries, extreme events can quickly wipe out hard-won development gains. In Mozambique annual disaster losses exceeded investment on three occasions between 1993 and 2011 (UNISDR, 2013).

The natural Nexus including climate also carries the potential for sudden, unexpected and unforeseen risks to global systems as all our existing weather is amplified. For example, extraordinary deep sea waves are now doubling in size from 20 metres to 40 metres high, threatening shipping (Wang *et al.*, 2014); this has the potential to disrupt the global food trade. When climate change enters the Nexus, "managed systems" can suddenly become unmanageable. As we become aware of the limits of human management, we need to analyze the Food–Energy–Water–Climate Nexus more carefully to identify its vulnerabilities, its weak links, in order to build resilience and strengthen our societies.

Which is the weakest link: Resource scarcity or disaster risk?

Within the anthropogenic Nexus, the conventional argument is to avoid or ameliorate resource scarcity in order to ensure food security, energy security and water security. However, as we have seen, there are large productivity gains still to be made in all three systems of the Nexus, and especially in water resources where governance is weak and fragmented, data is scarce and adequate pricing rarely applied. Furthermore, the anthropogenic Nexus is not the full story. Are we perhaps focusing too much on resource scarcity, at the risk of missing the bigger picture of the natural Nexus and its risks?

The costs of disasters are much higher than previously recognized: in the range of US$2.5 trillion so far this century (UNISDR, 2013). Of these costs, approximately 75 percent are due to water-related disasters. Looking further back, since 1900 88.5 percent of the thousand most disastrous events have been water-related. Therefore, in the natural Nexus water is a weak link within our interconnected and mutually dependent global systems. Strengthening this link, by managing water better to cope with increasing climate variability and extreme events, is a no-regrets strategy for climate adaptation. This involves paying much more attention to water storage, both natural and constructed, and both above ground and below. It also means better management of the way that water flows through river basins, bringing together all the relevant stakeholders, public and private, at community level, city level and national level, and ensuring that both water and risks are more equitably shared.

Another weak link in the natural Nexus relates to the lack of adequate climate information in the crucial time window between weather forecasts (up to about two weeks) and the detailed seasonal climate models (about six months ahead). For instance, the Russian heatwave and the massive Pakistani floods which both occurred in 2010 were only predictable nine days in advance (Williams and Simmons, 2014) – there was little "early warning" to enable disaster response mechanisms to react. This sub-seasonal gap in climate forecasting, between two weeks and six months, is the focus of a major global effort in developing the Global Framework for Climate Services (WHO, 2013). As climate variability intensifies, adequate early warning systems become ever more crucial to decision-makers, communities and economic actors in the Nexus.

Overall, the weakest link in the natural Nexus is clearly in the global energy system, which is still heavily and increasingly dependent upon fossil fuels such as coal, oil and gas. Until this is addressed, climate risks and the associated losses and damages will continue to mount. The most urgent and comprehensive risk reduction strategy of all is the decarbonization of the world economy, reducing carbon dioxide emissions to zero within the twenty-first century.

Disaster risk reduction as an economic and social driver of change

There are many potential economic opportunities to be seized in renewable energies, renewable water, agricultural adaptations such as drought- and salt-resistant crops, and in finding ways to decarbonize the global food system. Identifying efficiency gains in water–energy interlinkages can also generate cost savings, which pay back the necessary investments very quickly. However, let us also look at the potential for investing in disaster risk reduction, and especially in creating shared value between the private and the public sectors.

Reducing disaster risk requires getting the right signals to all the economic actors within the Nexus, especially the multiple private sector actors within the food and energy systems (such as farmers and power grid operators). However, a range of networks, communities and stakeholders need to be involved to bring about greater social awareness and change. Better climate information and robust early warning systems are essential, helping people and businesses to avoid or minimize economic loss and damage from disasters. With satellite imagery and mobile telecommunications, messages can be shared more quickly and effectively than ever before. It also makes sense to "build back better" – in rebuilding after a disaster, many adaptations can be made to reduce risks from future events.

Many more actors, including large insurers, are recognizing that investing in risk reduction can be much cheaper than the costs of the recovery effort, averting future costs and liabilities. Anticipating droughts and floods by putting pro-active policies in place requires good governance. A sustained investment on the part of governments, public sector actors and civil society in urban and rural areas is needed, in conjunction with private sector actors.

Building effective climate services and reducing risks requires both public and private sector involvement, linking multiple stakeholders including international organizations, national governments, research institutions and meteorological agencies, large corporations and small businesses in developing climate-smart approaches within the Nexus.

Managing the Nexus by strengthening the weakest links

The post-2015 agenda provides an excellent opportunity to strengthen the weakest links and create momentum for change in agreeing the Sustainable Development Goals. For the first time there is a dedicated goal and targets on water. In addition, in March 2015 the Sendai World Conference on Disaster Risk Reduction agreed a framework for disaster risk reduction 2015–2030, and in December 2015 the UN Framework Convention on Climate Change is set to agree a binding resolution on reducing carbon emissions. Clearly, the context for the Nexus will change. The evolution of the Nexus debate was predicated upon what I have called the anthropogenic Nexus – the Food–Energy–Water Nexus. These Nexus discussions have so far focused largely upon resource scarcity and the role of the private sector in meeting demand, increasing resource use efficiency, and recognizing the synergies to be gained in focusing on the interlinkages and interconnections of the three systems.

Broadening the Nexus to include climate involves the natural Nexus and the recognition of risk and unpredictability in these delicately balanced global systems. It requires a greater understanding and management of the changing global water cycle, not only water as a resource. Climate risk mapping changes the emphasis of the Food–Energy–Water–Climate Nexus, away from resource scarcity towards a broader approach to risk reduction. This in turn has implications for the relationships and linkages among the actors. There is a greater role for the public sector and for proactive policies in the natural Nexus – good governance is crucial. Reducing risks needs public and private sector actors to communicate and understand each other better, in real time.

In order to improve sustainability, avert catastrophic climate change in the future and reduce disaster risks now, our economies must decarbonize, and our societies must rapidly adapt to new levels of uncertainty. The water cycle in particular needs a stronger focus and better governance, for which the post-2015 agenda provides a historic opportunity. Both at international level and at national level, decision-makers need to pay more attention to the natural Nexus.

Natural resource security challenges for the Nexus

1. World food and feed production needs are projected to increase by 60 percent by 2050, although the world population is only set to grow by some 40 percent at most. The increase is due to projected changes in diet to include more meat, which in turn will mean more water, energy and feed requirements for greater numbers of livestock. While land as a resource is limited, there are large productivity gains to be made in food and feed production which will mitigate much of the additional resource requirements for water and energy.
2. World energy needs are also rising fast, if not as fast as food needs, since many technological systems are becoming increasingly energy efficient. At present, the increase in energy needs to 2050 is projected to rise in line with the world's population, by around 40 percent.
3. While some predict that there will be a water resource gap of 40 percent by 2030, this is unlikely. Increasing water efficiency and multiple reuse loops will keep water needs broadly in line with population growth (less than 40 percent by 2050) or even below.
4. The unpredictability of extreme climate events such as droughts and floods, and the economic loss and damage due to these disasters, are set to increase due to climate change, and they pose a greater security challenge to the Nexus than resource scarcity as such. Economic losses due to water-related disasters alone between 2001–2012 are estimated at US$2.5 trillion.
5. Decarbonizing our economies must be the top priority for resource security and sustainability.
6. Building greater resilience in the global food system and the global energy system is closely linked, via the natural Nexus, to strengthening the management of water resources as well as to reducing water-related disaster risks. At present, water risks from extreme events, rather than resource scarcity, are the weakest link in the Nexus.

References

Bahri, A., Drechsel, P., Scott, C.A., Raschid-Sally, L. and Redwood, M. (2010) *Wastewater irrigation and health*. London: Earthscan.

FAO (2012) *Coping with water scarcity: An action framework for agriculture and food security*. FAO Water Reports 1 38. Rome: FAO.

Faurès, J.-M., Svendsen, M. and Turral, H. (2007) 'Reinventing irrigation', in D. Molden (Ed.), *Water for food, water for life: A comprehensive assessment of water management in agriculture*, London: Earthscan, pp. 315–352.

ICOLD (2014) *Bulletin 149*. Available online at: www.icold-cigb.org/GB/Publications/bulletin.asp

IPCC (2011) *Climate change and water resources in system and sectors*. Available online at: https://www.ipcc.ch/pdf/technical-papers/ccw/chapter4.pdf

UNISDR (2013) *Global Assessment Report on Disaster Risk Reduction (GAR13): Creating shared value: the business case for disaster risk reduction*. Geneva: UNISDR.

UN-Water (2009) 'Global trends in water-related disasters, an insight for policy makers', in A. Adikari, and J. Yoshitani (Eds) *Global trends in water-related disasters: An insight for policymakers*, International Centre for Water Hazard and Risk Management. Available online at: http://unesdoc.unesco.org/images/0018/001817/181793E.pdf

Vorosmarty, C. (2014) *Water in the Anthropocene: New perspectives for global sustainability, GWSP*. Available online at: www.gwsp.org/products/current-opinion-in-environmental-sustainability.html

Wang, Z., Sorooshian, A., Prabhakar, G., Coggon, M.M. and Jonsson, H.H. (2014) *Impact of emissions from shipping, land, and the ocean on stratocumulus cloud water elemental composition during the 2011 E-PEACE field campaign*. Available online at: http://authors.library.caltech.edu/46158/

Williams, E.D. and Simmons, J.E. (2013) *Water in the energy industry: An introduction*. London, BP (British Petroleum) International Ltd.

World Health Organization (2013) *Global Framework for Climate Services*, Geneva: WHO. Available online at: http://www.gfcs-climate.org/sites/default/files/implementation-plan//GFCS-IMPLEMENTATION-PLAN-FINAL-14211_en.pdf

Zalasiewicz, J. *et al.* (2008) 'Are we now living in the Anthropocene?' *GSA Today*, 18(2), 4–8.

10 Natural resource security in an uncertain world

Sylvia Lee

Introduction

The rapid industrialization and economic growth of the twentieth century brought tremendous progress – life expectancy doubled and billions of people are no longer living in poverty. With this progress, however, new risks and vulnerabilities have also come to the fore. Exponential population growth and the burgeoning middle and urban classes are rapidly driving up the demand for water, food and energy, severely straining our natural resources and creating new sources of tension or conflict. Our hyper-connected economy and globalized supply chains have become more susceptible to vulnerability and shocks. At the same time, climate variability and change dramatically increase temperature, shift precipitation patterns and raise sea levels, adding additional layers of complexity and uncertainty.

Major challenges

Local and regional challenges become global catastrophes

Recent history has shown us that natural resource insecurity can destabilize a country or a region. In 2010, floods in Pakistan paralyzed large parts of the country for many weeks, killing thousands of people, severely reducing crop production and halving Pakistan's economic growth (Saeed, 2013). High and volatile international food prices in 2008 led to civil unrest in a number of cities in Africa, Asia and Latin America. Research has shown that volatile food prices lead to riots and contribute to political instability and intrastate conflict in low income countries, the places that are least responsible for changes in international food prices but which are most impacted (Lagi *et al.*, 2011). Conversely, the presence of a civil war is associated with an increase in domestic food prices (IMF, 2011). Similarly, recurring energy crises have sparked violent civil unrest, threatened the political survival of elected leaders, and contributed to economic and political inability to act in a number of developing countries. Paradoxically, attempts to improve energy security can also spark civil unrest. For example, the need for power in developing countries with untapped water resources is driving the construction of hydropower projects, which continue to create significant controversies and unrest (USAID, 2010).

The most concerning aspect of the resource insecurity challenge is how our interconnected world can transform a local and regional challenge into a global geopolitical and economic catastrophe. The central role that energy has played in transnational conflicts in the past several decades is well documented, particularly since the 1970s oil price spikes. Scholars estimate that up to half of interstate wars since 1973 have been linked to oil (Colgan, 2013). On the other hand, current literature cautions against focusing on 'food wars', 'climate wars' or 'water wars', especially since it is difficult to isolate one causal variable in complex conflicts. Some researchers have suggested that some of our contemporary armed conflicts are economic and socio-political in nature rather than climatological (Wischnath and Halvard, 2014) while others find consistent support for a causal association between climate events and conflict (Hsiang and Burke, 2013).

However, we do not yet understand the interconnectedness of these threats and how a stress in one of them may cause a tipping point in the threshold of a society's ability to respond. The increasing complexity in our food, energy, water and climate systems has multiplied the possible pathways for systemic global risks, and our globalized economy and supply chains are becoming more susceptible to shocks. The 2010 droughts in China, Russia and Ukraine and record rainfall in Canada reduced wheat production. As a response, governments imposed export bans that resulted in a doubling of global wheat prices and led to worldwide protests, which in turn sparked the start of the Arab Spring (Werrell and Femia, 2013). The local and regional droughts indirectly affected social unrest and created significant geopolitical impacts in the Middle East and North Africa. Another example is the 2011 floods in Thailand that caused hundreds of lives lost, millions of people displaced, and negatively impacted food prices and global manufacturing supply chains in the information technology and automobile sectors. The estimated US$45 billion direct and indirect economic loss places these floods among the costliest natural disasters on record (Grey *et al.*, 2013).

Difficult realities ahead

A number of scholars and policy experts have cautioned that these types of trends, where seemingly isolated local and regional threats become global catastrophes, will only get worse in the years to come. Demographic pressures and the growth of the middle class, particularly in developing countries, and its associated consumption are increasing global water, food and energy demands. Unless there are significant changes to the ways that we produce and consume, agricultural production will need to increase by about 70 percent by 2050, about 50 percent more primary energy will have to be made available by 2035 (Hoff, 2011), and the world will face a 40 percent shortfall between water supply and demand (Mckenzie and Company, 2009). Climate variability and change is increasing global temperature and variability of precipitation, making our water, food and energy resources more insecure, variable and uncertain.

Much of the current forecasts on water, food, energy and climate are focused on the biophysical systems and their related socio-economic impacts. Experts regularly conduct assessments and scenarios to paint a picture of what our future might look like, but most are sector specific. The World Water Council completes the World Water Development Report every three years, the International Energy Agency completes an annual World Energy Outlook, and the Food and Agriculture Organization publishes the biannual Food Outlook. The Intergovernmental Panel on Climate Change (IPCC) publishes its assessments once every five years to understand the scientific basis of the risk of human-induced climate change and its potential impacts and options for adaptation and mitigation.

However, few assessments exist that examine the interconnections between the different sectors. The United States' National Intelligence Council (NIC) cautioned in 2013 that there will be a significant risk of high and volatile international food and energy prices and it is highly likely that water shortages will continue to occur, which will increase the risk of discontinuous and systemic shocks (NIC, 2013). The most difficult to forecast or predict are the socio-economic and geopolitical impacts – but these have the largest potential systemic global significance in terms of natural resource insecurity risks. These types of low probability but high consequence global risks – from large-scale migration and conflict to global food market disruption and widespread panic – depend largely on how societies respond and how technologies advance, and they are the most difficult to predict and manage. These types of discontinuities and disruptions have the potential to fundamentally change the course of human history.

Inadequate systems and institutional arrangements

Today's market systems and institutional arrangements were mostly established in the past century, and are increasingly under strain in coping with the additional climatic and environmental stresses on our natural resources. State actors, single issue actors and the multilateral governance systems in place are ill-equipped to address increasingly complex, dynamic and fast moving threats, and they also have a difficult time incorporating new constituent voices, including the rise of important non-state actors.

As we move towards a more diffuse, multi-polar world order, many in the emerging economies are less and less content with the existing international institutions (e.g. multilateral development banks, the United Nations) set up after World War II. Countries such as Brazil, China and India have pushed for more say in international governance systems and have thus far been disappointed by the speed of reform. Slowly these countries are beginning to establish parallel regional and global institutions to rival existing ones. How these shifts in institutional systems will play out in the long term remains to be seen, but it is fair to assume that we cannot continue to rely solely on the institutions that are currently in place.

Moreover, observers often comment on how social media greatly facilitates the mobilization of informed citizens for political engagement and to participate in decision-making. Resources such as food and water are produced, consumed and managed by non-state or sub-national actors, not necessarily centrally managed by a central authority. The sheer number of actors and their differing preferences and actions further complicates decision-making (Jones, forthcoming).

While much of the de facto decision-making on climate, food, energy and water often happens at a local level, paradoxically someone acting in one country may have a direct impact in another country. Water resources flow downhill and do not respect political borders. The global market systems for food and energy trade move goods around the world, and a poor harvest in one location has an impact on the food supply in another. The global nature of climate change means that those least responsible for causing climate change may bear the brunt of its impact.

Finally, the interrelated nature of climate, water, food and energy means that one sector cannot tackle these challenges alone. Issues in one domain have direct consequences in another, and the causal connections require integrated analysis, a good understanding of the related trade-offs, and harmonized responses across sectors that go beyond simple coordination. Opportunities exist that generate additional benefits outweighing the transaction costs associated with stronger integration across sectors (Hoff, 2011).

Dealing with increasing uncertainty

The uncertainty that we face in relation to natural resource insecurity and its potential to cause global catastrophe paralyzes stakeholders and prevents meaningful action. Much has been written about the information gaps and the need for improved scientific understanding as it relates to climate change and its impact on water, energy and food, and the difficulties related to predicting extreme events. Similar to the warnings issued by financial institutions that past performance is not an indication of future results, in terms of projected changes in climate and resource volatility, historical data is not very useful for understanding the future (World Bank, 2012). Given the level of uncertainty that exists, it is not surprising that some climate models predict an increase in precipitation while others predict a decrease in various parts of the world for the same climate economic growth scenario (Turner and Annamalai, 2012).

These types of uncertainty become even more pronounced when longer-term projections are made and when interactions between resources need to be modelled. Compound effects mean that small adjustments in population or economic growth assumptions may have large consequences for resource projections (National Intelligence Council, 2013). Future emissions of greenhouse gases are greatly dependent on demographic and socio-economic trends, technological choices and policy decisions, which will have significant impacts on natural resources.

The complex and dynamic interplay between climate, food, energy and water insecurities, involving non-linear feedback loops and discontinuities, are also difficult to manage. At the global level, human activity has passed certain tipping points or planetary boundaries where there is a significant risk of irreversible and abrupt environmental change (Rockström *et al.*, 2009). At the societal response level, we do not know the threshold beyond which a drought may contribute to large-scale migration, nor do we understand how an abrupt escalation from smaller crises such as economic contraction and famines may have both resulted from and helped spark conflicts. These types of potential discontinuities contribute significantly to uncertainties.

Perhaps the most difficult uncertainty lies in how people and institutions will respond to increasing natural resource insecurity risks and increasing uncertainty in relation to climate variability and change. High prices and increasing volatility exposes vulnerability, and governments may choose to impose food export control and increase subsidies, as we have seen already during the 2010 drought (Werrell and Femia, 2013). Consumer panic and fear may also drive short-term and narrow self-interested behaviour, including hoarding. Large-scale migration threats may prompt countries to close their borders and prevent trade, or implement other radical policy changes. Individuals or groups may decide not to wait for governments to act on climate change and attempt to lower Earth's temperature through geoengineering and solar radiation management, which involves using particles in the atmosphere to reflect sunlight before it warms the planet, a strategy that will very likely have significant unforeseeable catastrophic environmental and societal side effects.

Potential solutions

Dealing with wicked problems

We first need to acknowledge that water insecurity, food insecurity, energy insecurity and climate change are all inherently 'wicked problems'. These are problems where it is very difficult, if not impossible, to measure or claim success because the problems bleed into each other (Rittel and Webber, 1973). Working at the nexus of these global challenges means that 'the answer' does not exist, but rather they are options – better or worse – that may improve certain situations. As the community grows to work on the Water–Food–Energy–Climate Nexus, we also need to accept that attempts to address these types of interconnected problems will have unintended consequences.

Strengthen systemic resilience

As a society, we need to strengthen our resilience enough to respond to the resource and environmental changes faster than the changes are happening.

Many resource-related conflicts are largely driven by the inability of states or communities to properly adjust to the speed of change. While we cannot prevent the resource volatility and variability from occurring and uncertainty will always remain, we can prevent catastrophic failure by strengthening resilience in our societies and our market systems. Fortunately, resource challenges are generally considered medium-term problems, which means that as a society we still have the time to mobilize our collective intellect and resources to strengthen our resilience (Jones, forthcoming).

For developing and emerging economies, we need to continue to alleviate poverty, strengthen economic systems, improve government effectiveness and the rule of law, and provide basic services to ensure long-term political stability. We also need to ensure that development, humanitarian and peacekeeping assistance continues where needed, and that equitable resource-sharing agreements are developed. Poor communities, and particularly women and marginalized groups, face the greatest peril. Many argue that the root causes of the Arab Spring were long-standing political grievances, mistrust, unemployment and economic marginalization. While the Arab Spring would likely have occurred one way or another, drought, food insecurity and climate change may have been the aggravating factors that made it arrive earlier (Werrell and Femia, 2013).

We need to develop plans and invest in ways that promote agile solutions, support redundancy, and create diversity in the sources of our water, food and fuel (NYS2100, 2013). Spare capacity or redundancy provides a measure of security when a water, food or energy system is under stress by ensuring that effective alternatives are available to respond to sudden or severe disruptions. Resilient networks are designed to prevent cascading systemic failures and allow for 'safe failure' that is limited in scope. When one domino falls it should not take down a whole system, so critical cross-sectoral independencies need to be highlighted and effectively managed.

Meaningful coordination between various sectors can happen more frequently by aligning and synchronizing incentives across sectors. Fortunately this is beginning to happen as some scientists, researchers, think tanks and governments are starting to embrace the Nexus approach. Since the Bonn conference in 2011, the Climate–Water–Energy–Food Nexus is slowly becoming a movement and is garnering greater attention and focus. Mechanisms need to be created to enhance integrated planning and develop criteria for integrated decision-making. The concept of a 'Chief Risk Officer' has been suggested for companies, cities, states and countries – a person or an office whose main responsibility is to ensure that systemic risks across a portfolio are managed, and who is in charge of cross-sectoral coordination (NYS2100, 2013; World Bank, 2014; World Economic Forum, 2007).

Finally, much has been written about how climate change is a threat multiplier (CNA, 2007). Since climate change and other deep uncertainties cannot be eliminated, flexible systems and frameworks need to be in place to help with decision-making. Flexibility is the ability to change, evolve and adopt alternative strategies in response to major disruptions, and recognize when it is not possible to return to the previous way things worked and find new solutions. This is particularly important when larger infrastructure investments are made that will have a very long life cycle, lasting for decades or even centuries with significant long-term consequences (World Bank, 2012). Some have called these 'no-regret' or 'low-regret' strategies – strategies that are likely to be good ideas even if predictions and forecasts are wrong. Some examples may include reducing water leakages in pipes, switching to crops more suitable to a warmer climate or improving energy efficiency.

Improving our predictive capacity under deep uncertainty

We must better understand how socio-economic pressures and climate and environmental change will impact our natural resources, and this will require much better data and information than we currently have. Innovations in mobile phones, computing power and data analytics give us the ability to have more information than ever. Advances in satellites and remote sensing allow us to observe natural resources from the sky in temporal and spatial granularity as never before, while advances in mobile technologies enable crowdsourcing of information. We are now at a moment in time where big data and crowdsourcing data can come together to provide us with a more complete picture of our resource and climate volatility. At the same time, technology also enables the diffusion of knowledge, which empowers greater transparency and information and allows researchers to build individual and community resilience. We are seeing shifts in how technologies such as mobile phones and social media can inform and mobilize citizens, and make decisions on issues that affect them directly.

We must also understand the complex web of conditions and interactions beyond climate, water, food, and energy – the larger system and the interactive effects of Nexus issues with politics, economics and social characteristics. While extremely challenging, the creation of a map of interacting complexity is necessary to understand the trade-offs and consequences of decision-making. We should also consider the use of tools such as scenario planning or war and security gaming to imagine the possible pathways and forecast how different responses may play out, since many of the outcomes depend largely on how governments and societies respond to increasing volatility.

To increase the likelihood that interconnectedness is mapped and understood, novel networks of diverse actors (e.g. physical and social scientists from multiple disciplines, trade experts, security experts, governments) are extraordinarily important. Looking back at how various predictions have fared in the past reveals that we are not very good at predictions. Predictions created in the 1970s for energy use in the US by the year 2000 were mostly wrong (Craig *et al.*, 2002). Academic research suggests that predicting events five years into the future is so difficult that most experts perform only marginally better than random, and in fact the wisdom of a set of diverse actors or a crowd is best at predicting the future (Horowitz and Tetlock, 2012).

The Water-Food-Energy-Security-Climate Nexus has consistently been identified as containing some of the greatest threats to the global economy and security (World Economic Forum, 2011). Resource insecurity is already acting as a threat multiplier in many parts of the world and will intensify considerably in the coming years. As the world crosses our environmental thresholds, unanticipated socio-economic impacts and frequent political instability will become the norm unless we strengthen our systemic resilience and improve our predictive capacity.

References

CNA Corporation (2007) 'National Security and the Threat of Climate Change', Alexandria: CNA Corporation. Available online at: https://www.cna.org/CNA_files/pdf/National%20Security%20and%20the%20Threat%20of%20Climate%20Change.pdf

Colgan, J.D. (2013) *Oil, Conflict, and US National Interests* Policy Brief, Belfer Center for Science and International Affairs, Harvard Kennedy School.

Craig, P.P., Gadgil, A. and Koomey, J.G. (2002) 'What can history teach us? A retrospective examination of long-term energy forecasts for the United States', *Annual Review of Energy and the Environment*, 27, 83–118.

Grey, D., Garrick, D., Blackmore, D., Kelman, J., Muller, M. and Sadoff, C. (2013) 'Water security in one blue planet: Twenty-first century policy challenges for science', *Phil. Trans. R. Soc. A*, 371, 20120406.

Hoff, H. (2011) *Understanding the Nexus*. Background Paper for the Bonn 2011 Conference: The Water, Energy and Food Security Nexus. Stockholm: Stockholm Environment Institute.

Horowitz, M. and Tetlock, P.E. (2012) *Trending Upward: How the intelligence community can better see into the future*. Available online at: www.ForeignPolicy.com

Hsiang, S.M. and Burke, M. (2013) 'Climate, conflict, and social stability: What does the evidence say?' *Climatic Change*, 123, 39–55.

International Monetary Fund (2011) *Food Prices and Political Instability*, prepared by Rabah Arezki and Markus Brückner, Washington, DC: IMF.

Jones, B. (forthcoming) 'Do Global Threats have Global Solutions? Reflections on the Governance of Transnational Issues', Washington, DC: The Brookings Institution.

Lagi, M., Bertrand, K.Z. and Bar-Yam, Y. (2011) *The Food Crises and Political Instability in North Africa and the Middle East*, Cambridge, Massachusetts: New England Complex Systems Institute.

Mckenzie and Company (2009) 'Charting our Water Future', Cicargo2030 Water Resource Group. Available online at: file:///C:/Users/Felix/Dropbox/nexus%202015/Charting_Our_Water_Future_Full_Report_.pdf

National Intelligence Council (2013) *Natural Resources in 2020, 2030, and 2040: Implications for the United States*. Available online at: www.dni.gov/files/documents/NICR%202013-05%20US%20Nat%20Resources%202020,%202030%202040.pdf

NYS2100 Commission (2013) 'Recommendations to Improve the Strength and Resilience of the Empire State's Infrastructure, New York: New York State'. Available online at: /www.governor.ny.gov/sites/governor.ny.gov/files/archive/assets/documents/NYS2100.pdf

Rittel, J.W.H. and Webber, M.M. (1973) 'Dilemmas in a General Theory of Planning', *Policy Sciences*, 4, 155–169.

Rockström, J., Steffen, W., Noone, K., Persson, Å., Chapin, F.S. III, Lambin, E.F., Lenton, T.M., Scheffer, M., Folke, C., Schellnhuber, H.J., Nykvist, B., de Wit, C.A., Hughes, T., van der Leeuw, S., Rodhe, H., Sörlin, S., Snyder, P.K., Costanza, R., Svedin, U., Falkenmark, M., Karlberg, L., Corell, R.W., Fabry, V.J., Hansen, J., Walker, B., Liverman, D., Richardson, K., Crutzen, P., and Foley, J.A. (2009) 'A safe operating space for humanity', *Nature*, 461, 472–475, doi: 10.1038/461472a.

Saeed, A. (2013) 'Floods have halved Pakistan's economic growth – expert.' *Thomas Reuters Foundation,* 9 September 2013.

Turner, A. and Annamalai, H. (2012) 'Climate change and the South Asian summer monsoon', *Nature Climate Change*, 2, 587–595.

US Agency for International Development (2010) *Energy Security and Conflict: A Country-Level Review of the Issues*, Washington, DC: USAID.

Werrell, C. and Femia, F. (2013) *The Arab Spring and Climate Change*, Washington, DC: Center for American Progress.

Wischnath, G. and Halvard, B. (2014) 'On climate variability and civil war in Asia', *Climatic Change*, 122, 709–721.

World Bank (2012) *Investment Decision Making under Deep Uncertainty,* Policy Research Working Paper 6193, Washington, DC: World Bank.

World Bank (2014) *World Development Report 2014 Risk and Opportunity: Managing Risk for Development*, Washington, DC: World Bank. Available online at: http://siteresources.worldbank.org/EXTNWDR2013/Resources/8258024-1352909193861/8936935-1356011448215/8986901-1380046989056/WDR-2014_Complete_Report.pdf

World Economic Forum (2007) Global Risks Report, Geneva: WEF. Available online at: www.weforum.org/pdf/CSI/Global_Risks_2007.pdf

World Economic Forum (2011) Global Risks Report, Geneva: WEF. Available online at: http://reports.weforum.org/global-risks-2011/

Part 4

Nexus perspectives

Energy: Water and climate

11 The Nexus in small island developing states

Liz Thompson

Background to the Nexus approach

The purpose of this chapter is to consider the benefit and possible approaches of a strategic formulation and implementation of "the Nexus approach" in SIDS.

In 2011 the Federal Government of Germany, joined by the European Union and a number of nongovernmental stakeholders including the World Economic Forum (WEF) and the World Wide Fund for Nature (WWF), initiated the first global conference on what was being called the Nexus of water, energy and climate. The Nexus approach emphasizes the interconnections between water, energy and climate, and more recently food. It also promotes more integrated and holistic approaches in policy and programming and a shift toward "sustainable consumption and production." At the heart of the 2011 Bonn Conference was the idea that "a nexus perspective helps to bring economic benefits through more efficient utilization of resources, productivity gains and reduced waste."

The central argument in favour of the Nexus approach is the contention that "failing to recognize the consequences of one sector on another can lead to notable inefficiencies in the system." Proponents of the Nexus urge integrated, coherent policy and planning in the cluster of thematic areas because "decisions on the type of energy generation can significantly influence water demand and in the case of biofuels, compete over land for food production; the way water is sourced, treated, priced and distributed can raise or lower energy requirements; and the choices made on food and diet influence both water and energy needs."

Since Bonn, the Nexus approach has generated dialogue, academic papers and further conferences, all of which have sought to advance the Nexus as being at the core of the interconnection between society, economy and ecology, the three pillars of sustainable development. The Nexus has been described as the thread which sews together the three pillars of development and "the means by which the dots are connected." This is an important argument, since in some quarters there is a school of thought which is of the opinion that describing development as having three pillars fails to visually convey their interconnectedness. It has also been argued that the triple helix concept, analogous to a strand of DNA, more effectively conveys their interlinkage and interdependency. Despite the presence of these arguments in the policy sphere, the "three pillars of sustainable development" remains the preferred universal expression.

In the intervening period since 2011 academics and practitioners in development have also pointed to other areas of the Nexus, such as water, energy and women's health.[1] Four clear thematic areas have now emerged in the Nexus dialogue – water, energy, food and climate.[2] These were the focus of the University of North Carolina's Nexus Conference in 2014. Each theme separately and collectively represents critical components of sustainable development. The UNEP has defined a green economy as one that results in "improved human well-being and social equity, while significantly reducing environmental risks and ecological scarcities" (UNEP, 2010). The Nexus is pivotal to the organization of a green economy, which grows wealth and builds human and social capital with greater efficiency while simultaneously protecting the natural resource base. Policy, planning and implementation of a green economy exhibit a higher degree of coherence and integration and eschew the silo approach. The Nexus approach uses and generates greater efficiencies, maximizes resources and reduces waste "without compromising sustainability and exceeding environmental tipping points."

Sustainable development can be thought of as a wall still under construction, with the bricks made of myriad thematic components and the Nexus serving as the mortar which holds them all together, necessary to erect the whole and to make it cohesive and impregnable.

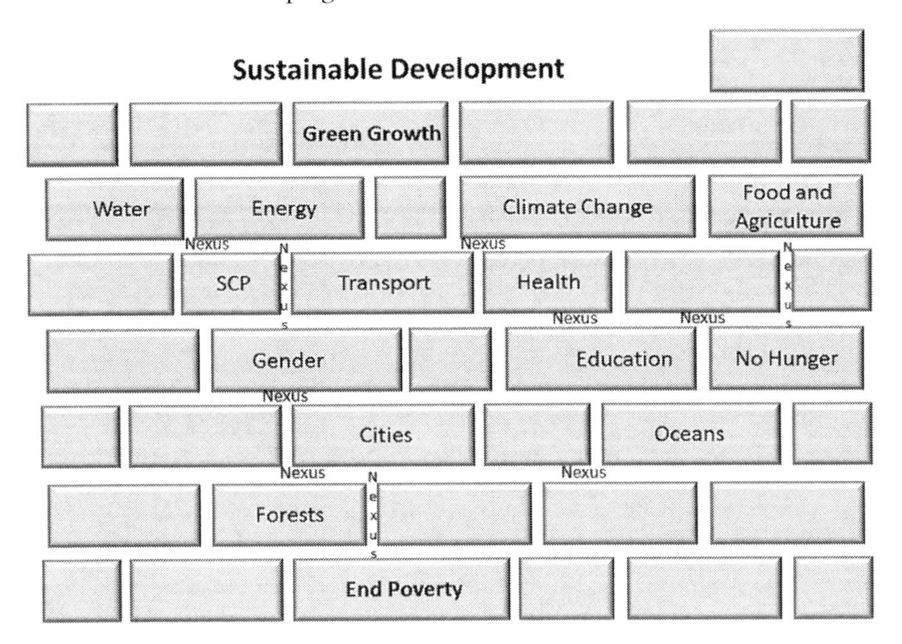

Figure 11.1 Sustainable development as a wall under construction.

A number of social, economic and environmental factors such as resource scarcities, an expanding global population, bust and boom economic cycles, the need to raise quality of life for the "bottom billion" who live in abject poverty and deprivation and the severe global impact of climate change, are fuelling greater emphasis on greening the global economy. This thrust includes natural resource protection and more sustainable lifestyles and production. Some insist that natural resources should be valued, measured and accounted for in economic activity and indices which should go beyond the conventional measurement of gross domestic product (GDP). Consequently there has been an emergence of "green accounting systems" such as WAVES – Wealth Accounting and the Valuation of Ecosystem Services; TEEB – The Economics of Ecosystems and Biodiversity; and SEEA – the System for Environmental Economic Accounting. As these new accounting systems become more utilized and entrenched, greater emphasis will be placed on natural resources, how they are used and their contribution to building wealth and social capital. Increasingly, therefore, Nexus thinking, even if called by another name, will become a valuable tool in economic discourse, wealth assessments and policy approaches.

Despite this apparent logic and usefulness, the Nexus approach has to date been sparingly applied. The Nexus was not part of early development platforms such as Agenda 21,[3] the BPOA[4] and the MDGs.[5] Although they all include the Nexus themes of water, energy, food and climate, Nexus thinking has not found its way into the global development platforms that have been erected since the Bonn Nexus Conference, such as the Future We Want,[6] the SAMOA Pathway[7] and the SDGs.[8] However, at the national level in SIDS Nexus approaches are being adopted, even if they are not being so-called.

The global circumstances

It is necessary to understand the background against which Nexus thinking emerged. The FAO informs us that a billion people worldwide are undernourished or go hungry. The International Energy Agency estimates that in 2011, 1.3 billion members of the global family did not have access to energy and 2.6 billion were reliant on biomass, including wood, animal or crop waste, for cooking. The World Health Organization indicates that 50 percent of premature deaths in children under five are caused by "household pollution", which is also responsible for 4.3 million annual deaths globally, with the majority of those deaths occurring among women and children. UNEP figures suggest that a further 0.9 billion lack access to safe drinking water, while 2.6 billion are without proper sanitation. Climate change is regarded as a significant development challenge with a causal link to greenhouse gas emissions from the extraction, production and consumption of fossil fuels. Research by the DARA Group reveals that the adverse impacts of climate change have a global cost of US$1.2 trillion per annum, wipe 1.6 percent per annum from global GDP, and result in 400,000 deaths.

The Nexus not only considers the contribution of these elements to development and conversely the contribution made by their absence to underdevelopment and poverty, but also considers how they should be collectively pursued for improved and sustainable development outcomes. Energy is used in the extraction, treatment, desalination and distribution of water, sewage disposal and sanitation. Water is essential to every stage of energy production including extraction, processing, distribution, hydraulic fracturing (fracking), the cooling of generating plants and systems, sewage disposal and sanitation. Fracking is becoming a more popular means of oil and gas production and uses about six million gallons of water each time a well is fracked.

The UN World Water Development Report (2014) puts it this way:

> Water and energy are tightly interlinked and highly interdependent. Choices made in one domain have direct and indirect consequences on the other, positive or negative. The form of energy production being pursued determines the amount of water required to produce that energy. At the same time, the availability and allocation of freshwater resources determine how much (or how little) water can be secured for energy production. Decisions made for water use and management and for energy production can have significant, multifaceted and broad-reaching impacts on each other – and these impacts often carry a mix of both positive and negative repercussions.

The majority of energy used globally derives either directly or indirectly from fossil fuels, accounting for 87 percent of primary energy use in 2012. In Caribbean SIDS particularly, the national water authority/company is usually the largest single client of the utility company or power generator. The International Energy Agency estimates that across the globe 580 billion cubic metres of water are extracted annually and used in energy-related processes. The Agency estimates that this amounts to 15 percent of global water withdrawal and is second only to the volume of water extracted for use in agriculture. Both energy and water are required for food production. Fossil fuels are a significant contributor to carbon emissions, which cause climate change impacts.

The global population, which now stands at seven billion, is expected to grow to nine billion by 2050. The IEA is projecting that global energy demand will rise by 0.7 percent to 1.4 percent per annum through to 2035. The World Energy Council anticipates that the water consumed in energy processes will increase by 10 percent to 18 percent through to 2050. This raises a clear issue about the need for clean air and protected ecosystems, together with the demand of the global population for energy and water for life and development, versus competing and conflicting resource uses (commercial, industrial, recreational). As global populations grow, urbanization increases, more people are taken out of poverty, demands for natural resources and food increase, and so too does the demand for water and energy. This trend calls for more sustainable and efficient extraction and production methodologies. Pressure is increasing to generate energy from more sustainable resources, to conserve water, to ensure greater natural resource efficiency, and to minimize climate impacts.

Small island developing states (SIDS)

There are fifty-one small island developing states in the geographic regions of the AIMS,[9] the Caribbean and the Pacific. The world's small island developing states range in size from Niue at 100.4 square miles (260 square kilometres) to Cuba at 42,426 square miles (109,884 square kilometres). Singapore, which at number 25 is the highest-ranking SIDS of the 187 countries on the 2013 Human Development Index, has a GDP of US$297.9 billion, and a per capita GDP of US$55,182.48. Guinea Bissau at number 176 is the lowest ranking SIDS; its GDP is US$858.7 million and its per capita GDP is US$503.83. The largest SIDS population can be found in Cuba, with 11.3 million people in comparison with Niue's 1500, the smallest population. The single most common and largest foreign exchange earner in SIDS is tourism. Islands in the Caribbean, AIMS and the Pacific are among the world's most tourism-dependent economies,[10] and are significantly impacted by seasonality and volatility in this industry. Bahrain and the Caribbean also earn substantial revenues from banking and financial services. Trinidad and Tobago's economy is driven by oil and gas exports. The economies of Pacific SIDS are dependent on tourism, agriculture and fisheries, and to a lesser extent so are those of Eastern Caribbean countries.

The current global development agenda rests on a platform known as Agenda 21, which was birthed at the first Rio Conference in 1994. The year 2015 fast approaches and with it the roll out of the "post-2015 development agenda" on which the emerging global development agenda will be grounded from 2015. This changing global landscape has resulted in an expanded and intense dialogue on the most effective approaches to development. The Barbados Programme Of Action (BPOA) for Small Island Developing States created some fourteen strategic priorities for SIDS. These included climate change, natural and environmental disasters, land degradation, and marine and natural resources. The further and most recent articulation of the SIDS development agenda came at the Third International Conference on SIDS held in Samoa in September 2014. The resultant Outcome Document, "The SAMOA Pathway", addresses a list of twenty-nine strategic priority areas, including those of the Nexus – "climate change, sustainable energy, food security and nutrition, water and sanitation" – although the SAMOA Pathway does not specifically advocate a Nexus approach.

The circumstances of SIDS

The 1994 SIDS Conference acknowledged that SIDS have the rights of sovereignty, self-governance and to determine their own development agenda. It noted that SIDS have peculiar development challenges which are exacerbated by their small size, isolation, distance from centres of production, high population density relative to small size, limited human resource capacities, inability to create economies of scale and scope, and vulnerability to environmental and exogenous economic shocks. Many SIDS are highly indebted with high debt-to-GDP ratios. Some, like Barbados, Jamaica, St Kitts and Singapore, have a debt-to-GDP ratio that is over 100 percent. These factors, along with "disproportionately expensive public administrations", are regarded as the defining characteristics of SIDS, which impact on their overall sustainability as well as the development of their societies, economies and environments.

Size and isolation further impact trade opportunities – islands are unable to produce large volumes at competitive prices, and distance makes shipping and related costs prohibitive. The United Nation's Department of Economic and Social Affairs (DESA) advances the view that "these special vulnerabilities SIDS face accentuate other challenges facing developing countries in general. These include, among others: difficulties in benefitting from trade liberalization and globalization; heavy dependence on coastal and marine resources for their livelihood including food security; heavy dependence on tourism which can be easily impacted by climate change and natural disasters; energy dependence and access issue; the limited freshwater resources; limited land resulting in land degradation, which affects waste management, and vulnerable biodiversity resources."

Water in SIDS[11]

In 2013 the World Resources Institute (WRI) produced a list of the world's most water-stressed countries.[12] Ten SIDS came within the top twenty countries on the list with the highest possible stress score of 5.

Table 11.1 The world's most water-stressed countries.

Top countries facing water stress			
Rank	*Country name*	*Baseline water stress score*	*SIDS*
1	Antigua and Barbuda	5.00	X
2	Bahrain	5.00	X
3	Barbados	5.00	X
4	Comoros	5.00	X
5	Cyprus	5.00	
6	Dominica	5.00	X
7	Jamaica	5.00	X
8	Malta	5.00	*
9	Qatar	5.00	
10	Saint Lucia	5.00	X
11	Saint Vincent and the Grenadines	5.00	X
12	San Marino	5.00	
13	Singapore	5.00	X
14	Trinidad and Tobago	5.00	X
15	United Arab Emirates	5.00	

Top countries facing water stress			
Rank	*Country name*	*Baseline water stress score*	*SIDS*
16	Western Sahara	5.00	
17	Saudi Arabia	4.99	
18	Kuwait	4.96	
19	Oman	4.91	
20	Libya	4.84	
21	Israel	4.83	
22	Kyrgyzstan	4.82	
23	East Timor	4.81	
24	Iran	4.78	
25	Yemen	4.67	
26	Palestine	4.63	
27	Jordan	4.59	
28	Lebanon	4.54	
29	Somaliland	4.38	
30	Uzbekistan	4.32	
31	Pakistan	4.31	
32	Turkmenistan	4.30	
33	Morocco	4.24	
34	Mongolia	4.05	
35	Kazakhstan	4.02	
36	Afghanistan	4.01	

* Malta is a developed country. X = SIDS with a score of 5

Energy in SIDS

Small island developing states base some 90 percent of economic and social activity on energy derived from imported fossil fuels. This puts them at considerable economic risk. SIDS are located far from the centres of fossil fuel extraction and production and have high freighting costs. Their small sizes and population bases also act as a barrier against competitive volume pricing; the creation of economies of scale makes profit generation for power producers and distributors difficult. With the exception of Trinidad and Tobago which is a mature producer of oil and gas, the vast majority of island states are highly vulnerable to oil price fluctuations and are hit particularly hard when oil prices rise.

High oil prices and their impact on SIDS present a major risk to their social and economic security. Prior to the historic spike in global oil prices to US$147 per barrel in July 2008, the UNDP estimated that SIDS were already collectively spending US$67 million per day on oil importation, with the average oil import bill "in the Caribbean region 21 percent of GDP; in the Pacific SIDS it was 18 percent of GDP." Many island states are locked into contractual monopolies with utility companies, which are diesel-using generators. Existing contractual relationships therefore militate against the use of new and energy efficient technologies, exploration of off-grid solutions and shared power from renewable energy sources. As a consequence of these combined factors, the energy prices in SIDS are extraordinarily high.

Climate change in SIDS

SIDS populations collectively contribute less than 1 percent of total greenhouse gas emissions, but they are disproportionately impacted by this phenomenon with its most extreme impacts evident in the sinking of the Maldives and the rising intensity and frequency of extreme weather events, growing water stress and scarcity, desertification, loss of coastlines, adverse impacts on marine habitats and warming and rising sea levels, amongst other things. Expenditure on climate change mitigation and adaptation in SIDS is as high as it is inescapable, and is very much a part of the cost of development and sustainability in island states.

Food and agriculture in SIDS

SIDS grow very little of the food they consume. In 2006 the Caribbean's food import bill was US$5 billion for a tourism-based economy and a resident population of 35 million people. The Food and Agriculture Organization (FAO) reports: "Globally, in 2008, the food import bill for low-income food-deficit countries increased by about 35 percent from 2006."[13] The FAO has pointed to food imports accounting for 21 percent of the total regional import bill in the Caribbean.

The SIDS island trap

Collier (2007) advances the idea that there are four traps that keep countries in poverty and their populations languishing "in the bottom billion" of the development spectrum. He identifies these as conflict, natural resources, being landlocked with bad neighbours, and bad governance in a small country. While scholars have spoken of the special characteristics and vulnerabilities of SIDS, it can be argued that these very features act as a trap just as effectively, constraining growth and economic expansion in SIDS. Those who seek to counter this idea of small island developing states being trapped by their size, isolation and other peculiar characteristics may point to the fact that a majority of SIDS have graduated from LDC status, and many are ranked as middle- and high-income countries. While that is true, this argument fails to recognize the fact that despite some growth, even SIDS in the high income bracket are heavily indebted, many in excess of 100 percent of GDP, and their vulnerability to external shocks creates high risk for jeopardizing or reversing progress and becoming sustainable.

The Nexus approach in SIDS

The Rio Conference of 1992 served to catalyze greater interest in environmental stewardship and policy development. It also gave impetus to sustainable development as a discipline and a desirable and attainable objective for the global family of nations. Significant progress has been made in this regard, but sustainability and sustainable development have not become an embedded (mainstreamed) principle and governing ethic by which policy is formulated, business is conducted or daily activities are pursued, despite the potential efficiency and development gains of such an approach. Embedding sustainability is about applying Nexus thinking and practices across government, business and society.

Limited resources and populations, high costs and debt, peculiar defining characteristics and vulnerabilities would make the Nexus approach in SIDS particularly beneficial in embedding and achieving sustainability. The Nexus is now finding its way into policy planning and implementation in SIDS, with a number of deliberate Nexus policy choices and other initiatives that span the gamut of the Nexus thematic areas. However, it is important for governments to take the lead in creating the enabling policy, legal, governance, regulatory, incentives and general environment for capital to enter and invest in markets with confidence, and in promoting innovative technologies which will be impactful in the Nexus and in general sustainability. Greenspan (2007) argues that it is the ultimate responsibility of the government to create the environment for innovation and investment, and this is especially true in relation to the Nexus.

- Barbados developed a Green Economy policy in 2007, which seeks to use integrated policy approaches which are Nexus-oriented. A similar policy approach is being discussed in Trinidad and Tobago.
- Ocean Thermal Energy Conversion (OTEC) is still an expensive, experimental technology considered suitable for tropical and subtropical climates. It has the potential to bridge all areas of the Nexus since it generates energy from a renewable source (the solar energy absorbed by the ocean) and produces cool water for air-conditioning systems, irrigation and greenhouses, nutrient-rich water for fish farming, and fresh water. OTEC pilot systems are currently being considered in the Bahamas, Barbados, Curacao, Jamaica, Martinique and Puerto Rico.
- St Kitts and Barbados are currently considering waste-to-energy facilities. This technology also addresses the areas of the Nexus. Energy generated in this way could be used for numerous activities.
- Aruba is moving to 100 percent renewable energy, and in so doing will pursue a number of green economy and Nexus approaches.
- St Lucia, Grenada and Dominica are moving toward more sustainable practices, starting with energy and water conservation.

- Tokelau in the Pacific, which comprises three small atolls administered by New Zealand, has transitioned to 100 percent renewable energy using solar and biofuel from coconuts.
- Nauru, Palau and Vanuatu are investing in solar-powered desalination plants.
- Jamaica is already using ethanol from sugar cane as a transport fuel, thereby reducing its carbon emissions. The benefit of using sugar cane is that it is a grass, which uses little water and aids in soil retention. It is a clean fuel, it reduces carbon emissions, and it involves the Nexus issues of water, energy and climate change. Insofar as it does not take food stock away from the market in the way that corn might, it may be said to positively impact a third Nexus issue, food and food security.
- Barbados developed an indigenous solar water-heating industry in the 1970s and now has the fifth highest per capita penetration in the world. Barbados' solar water heaters (SWH) currently save the country from importing 185,000 barrels of oil or 200 million kWh annually. The Central Bank of Barbados estimates that between 1975 and 2002,[14] with 35,000 SWHs installed, Barbadian consumers saved US$130 million[15] on energy expenditure and avoided 15,000 metric tonnes of carbon emissions.
- Specific strategies for the use of grey water for irrigation, as well as the nutrient-rich tertiary treated water from sewage, could be part of a Nexus approach in SIDS, which as tourism destinations frequently irrigate golf courses and hotel grounds with potable water.
- SIDS from every geographic region of the world endorsed and committed to Sustainable Energy for All (SE4All) in the Barbados Declaration in 2012. The SE4All initiative promotes achieving universal access to modern (renewable and sustainable) energy services, doubling the rate of renewable energy technologies and usage and doubling the current rate and level of energy efficiency, all by 2030. The application of SE4All in the national energy frameworks of SIDS will address four of the Nexus areas: climate, energy, water, and food.

The development trajectory of SIDS will be determined by how well island states address issues of water, energy, climate and food security. Islands must shift from resource-wasteful technologies and practices to integrating the Nexus in policy planning and programming, in order to maximize resources and opportunities for efficiencies, cost effectiveness and growth. Further strategies should include assessing natural capital, utilizing renewable energy resources, using size and scalability in SIDS as an advantage for the creation of models of green development and sustainability, deploying regional approaches to create economies of scale and scope, developing agricultural systems which use low energy and water, promoting government policies and initiatives which reflect the need for coherent, coordinated approaches in Nexus areas, integrating water, energy and food production into national and regional climate strategies, incentivizing the local private sector toward green innovation and sustainability practices, and mobilizing resources for Nexus initiatives and broad development planning.

Financing development in SIDS

In UNFCCC negotiations, SIDS have sought to achieve agreement about keeping carbon driven global temperatures at 1.5° Celsius, rather than the compromise of 2° Celsius. SIDS bear what has been described as a "disproportionate burden" of climate change impacts, which compels their policy makers to focus their attention on adaptation, mitigation and financing measures to address warming sea temperatures, rising levels, decreasing pH, coastal damage and inundation. Many, like the Maldives, are disappearing, while all SIDS are experiencing increased frequency and intensity of extreme weather events. In 2004 Hurricane Ivan, a category 5 hurricane, hit Grenada and took only a few hours to wipe out 90 percent of the island's housing and 200 percent of its GDP. The level of climate change financing, adaptation and mitigation costs results in significant need for a SIDS-specific carbon-financing instrument which takes account of their natural capital.

Oceans and seas cover 70 percent of Earth's surface and are essential to the health of the planet's ecosystems. They are important sources of food, biodiversity and minerals, as well as potential sources of energy. The marine environment also plays an important role in the tourism economies of SIDS. In light of evolving global, green development metrics, what tools would best evaluate SIDS national patrimony, natural resources and marine potential in the context of providing capital to facilitate their climate protection, finance energy, water and agricultural systems and the general development prospects of island states?

Through new partnerships between governments and businesses such as PPPs, BOLTs and other similar arrangements, private sector capital is playing a greater role in financing government and development projects, both globally as well as in SIDS. To enable access to new sources of financing, natural capital assessments in SIDS must take into account the maritime resources of islands, which greatly surpass their terrestrial resources. Some examples will make this point clear. In the Pacific, Tuvalu has a land area of 26 square kilometres or 10.04 square miles. Its exclusive economic zone (EEZ) is 751,797 square kilometres, or 467,144 square miles. Samoa's land area measures 1137 square miles or 2944 square kilometres. Its EEZ is 131,812 square kilometres or 81,904 square miles. In the Caribbean the Bahamas, which have a coastline of 11,238 kilometres or 6,385 miles, lay claim to an EEZ of 369,149 square kilometres or 142,529 square miles. Barbados, with a coastline of some 97 kilometres or 66 miles and a total landmass of 432 square kilometres (166 square miles), records an EEZ of 183,436 square kilometres or 113,981 square miles.

The REDD+ initiative was developed to finance and support activities which reduce carbon emissions and militate against deforestation and degradation in countries with substantial forest cover. Seas and oceans, like forests, are effective carbon sinks. This raises a couple of questions:

- Could not a similar marine mechanism be developed for SIDS, particularly with regard to their large maritime territory?
- What would the structure of this mechanism look like?

Such a mechanism would be important to natural resource accounting in SIDS as well as in evolving environmental and development metrics. It would also be useful in the mobilization of finance for research and activities related to climate change adaptation and mitigation, poverty eradication, sustainable human development, and in effecting transition to sustainable energy and a green economy. The mechanism would be compatible with provisions in the UNFCCC and the Kyoto Protocol for developed country parties and other parties included in Annex II to the Convention, to support technology transfer, provide new and additional financial resources, and support activities and projects which address climate change impacts, including the protection of sinks in developing countries.

It is this author's suggestion that the mechanism's structure would include:

- A Board with representation from SIDS regions, their representative organizations, development partners, and relevant UN, international and technical agencies.
- A Secretariat staffed by multidisciplinary experts.
- A Blue Investment Fund or finance instrument to provide SIDS with the financial resources to manage and protect marine resources, ecosystems and coastlines.
- A Development Fund to provide financing for development activities and Nexus policies and projects in energy, water, food security, climate protection and disaster risk reduction.

The mechanism would have:

- Capacity for implementation, monitoring, measurement, reporting and verification.
- Provisions to protect the value of oceans and seas as the environmental lifeblood of SIDS and the planet.
- Capacity to track and monitor the health of seas and oceans and their effectiveness as sinks.
- Capacity to inventory and evaluate marine flora and fauna, including but not limited to coral reefs, kelp, sea grass beds and minerals.
- Knowledge platforms for sharing technology, research, best practices, data and information for decision making.
- Compatibility with UNFCCC and its instruments including the Kyoto Protocol.

The mechanism would enhance SIDS':

- Capacity for partnership and collaboration with international development and financial institutions.
- Potential for complementary funding under the UNFCCC including the CDM, GCF, GEF and existing and new financing mechanisms.
- Opportunities for emissions trading schemes and funding with Annex II Parties and I.
- Potential to maximize evolving green and non-GDP measurements such as TEEB, SEEA and WAVES.
- Delimitation of maritime boundaries with corollary assessments of the contribution of EEZs to national patrimony and wealth.

Conclusions

The development potential of SIDS is constrained by their size and peculiar characteristics. "The Samoa conference suggested that SIDS could … become bellwethers for sustainable development, modeling whole-of-society approaches to renewable energy, waste management and other aspects of sustainability" (IISD, 2014). The Nexus approach emphasizes the interconnections between common development challenges and common development solutions. The Nexus also provides the opportunity for resource optimization, greater efficiencies and more strategic targeting of development objectives across all sectors. It is now up to the governments of SIDS to work more collaboratively across geographical boundaries, using Nexus thinking and approaches to achieve development objectives, poverty eradication and sustainability.

Notes

1 Sustainable Energy for All (SE4All) makes this point. Website text: www.se4all.org/hio/energy-and-womens-health/
2 Nexus 2014: Water, Food, Climate and Energy. Website text: http://nexusconference.web.unc.edu
3 The development platform birthed at the Rio Conference on Sustainable Development 1992.
4 Barbados Programme of Action for Small Island Developing States (SIDS) 1994.
5 The Millennium Development Goals of 2000, the brainchild of the development institutions UN World Bank.
6 The Future We Want – The Outcome Document of the Rio+20 Conference on Sustainable Development 2012.
7 SIDS Accelerated Modalities Of Action Pathway – The Outcome Document of the Third International Conference on SIDS 2014.
8 Sustainable Development Goals which are at the heart of the global development agenda for 2015 and beyond.
9 African, Indian Ocean, Mediterranean, South Pacific Sea.
10 Guyana is a member of the Caribbean Community and is 83,000 square miles/214,970 square kilometres.

11 Small Island Developing States membership: http://sustainabledevelopment.un.org/index.php?menu=1520
12 Sustainable Development Knowledge Platform, Small Islands Developing States: http://sustainabledevelopment.un.org/index.php?menu=203
13 Water stress is measured by a score of 1 (lowest stress level) to 5 (the highest).
14 Recent and comprehensive data are available for the Caribbean and other SIDS.
15 It is difficult to obtain more current figures as the installation level of SWHs is not generally tracked.

References

Barbados Declaration 2012, available online at: http://www.undp.org/content/dam/undp/library/Environment percent20and percent20Energy/Climate percent20Change/Barbados-Declaration-2012.pdf

Bonn 2011 Conference Water Energy and Food Security Nexus – Solutions for a Green Economy 13 February 2012 Policy Recommendations, available online at: www.water-energy-food.org/.../bonn2011_policyrecommendations.pdf

Briguglio, L. (1995) 'Small Island Developing States and their Economic Development', *World Development*, 23(9), 1615–1632.

Collier, P. (2007) *The Bottom Billion*, Oxford: Oxford University Press.

DARA Group and Climate Vulnerable Forum (2012) *Climate Vulnerability Monitor: A Guide of the Cold Calculus of A Hot Planet*, available online at: http://daraint.org/wp-content/uploads/2012/09/CVM2ndEd-FrontMatter.pdf

Greenspan, A. (2007) *The Age of Turbulence: Adventures in a New World*, London: Penguin Books.

IISD (2014) *Policy Update: Small Island Developing States: Shaping the Sustainable Development Agenda in a Terrestrially-Focused World*, Winnipeg: IISD. Available online at: http://sd.iisd.org/policy-updates/small-island-developing-states-shaping-the-sustainable-development-agenda-in-a-terrestrially-focused-world/

Nexus Conference (n.d.), available online at: http://www.water-energy-food.org/en/whats_the_nexus/bonn_nexus_conference.html

Nile Basin Discourse (n.d.) *One Nile One Family – Water, Food and Energy Security in a Changing Climate*, available online at: www.nilebasindiscourse.org

UNEP (2010) *Green Economy Developing Countries Success Stories*, Geneva: UNEP.

United Nations World Water Development Report (2014) *Water and Energy* Volume 1, Paris: UNESCO. Available online at: http://unesdoc.unesco.org/images/0022/002257/225741e.pdf

Water, Energy and Food Security Resource Platform (n.d.) *What's The NEXUS: Messages and Policy Papers*, available online at: http://www.water-energy-food.org/en/whats_the_nexus/messages_policy_recommendations.html

12 Renewable energy

Nexus-friendly pathways for growth

Frank Wouters and Divyam Nagpal

Introduction

The water, energy and food nexus is at the centre of global policy, development and research agendas. The rise in its prominence has been spurred by the prospect of meeting rapidly growing demand in an increasingly resource-constrained world.

Meeting growing demands for water, energy and food is becoming progressively more challenging as resource constraints become more pronounced and external factors, such as climate change, limit the ability of existing systems to expand capacity. Between now and 2050 energy and food demand will nearly double, with water demand increasing by 55 percent (OECD, 2012).

The interlinkages between the three systems will grow stronger. By 2035 water withdrawal for energy is projected to rise by 20 percent, with consumption increasing by 85 percent (World Bank, 2013). Similarly, water needs for agriculture and domestic purposes will increasingly be met through resources that are harder to reach and more energy-intensive to exploit. Producing more food will require land, water and energy inputs, with potential trade-offs involving the increasing use of bioenergy.

Renewable energy technologies can be an optimal solution to reduce strains on specific elements of the Nexus and enhance security. The rapid adoption of renewables over the past decade has been driven by ambitions to improve energy security, expand energy access, mitigate climate change and stimulate socio-economic development. The distributed, sustainable, environment-friendly and less resource consumptive nature of renewable energy means that they can also reduce strains on specific elements of the Nexus. They can further sustainable development objectives and assist governments in optimizing resource security across all sectors. However, quantitative and qualitative knowledge on the role renewables can play in addressing Nexus-related challenges remains dispersed and limited.

As a first step towards bridging the knowledge gap, IRENA undertook a study entitled '*Renewable energy and the Nexus*' to bring to the forefront the opportunities presented by renewables. The report adopted a life-cycle approach to identifying renewables intervention in the water and food sector. This chapter presents some of the key findings from the study.

Renewable energy and the food–energy Nexus

The food sector accounts for nearly a third of global energy consumption (FAO, 2011). Food production involves diverse flows of energy. As illustrated in figure 12.1 the majority of the energy inputs are at the processing, distribution, retail, preparation and cooking stages, when compared to production. Variations remain between high-GDP and low-GDP countries given differences in production practices, the extent of mechanization and efficiencies of processing and distribution infrastructure.

High dependence of the food sector on fossil fuels offers a real threat to the long-term viability of food security.[1] Heavy reliance of food systems on fossil fuels can negatively impact the continuity of physical and economic access to food products. For instance, the 2007–08 global food crisis was a direct consequence of increasing oil prices, which had a cascading effect, leading to a greater demand for biofuels and trade shocks in the food market (EU ERC, 2011).

Enhancing the role of renewable energy within food systems allows decoupling from fossil fuel energy, thereby contributing to increasing food security. Renewable energy can be used throughout the food sector either directly to provide energy on-site or indirectly by integrating this energy into the existing conventional energy supply. Renewable energy is feasible for farm and aquaculture production, as well as for transporting raw food feedstock, processing food, distributing finished products and cooking (FAO, 2011).

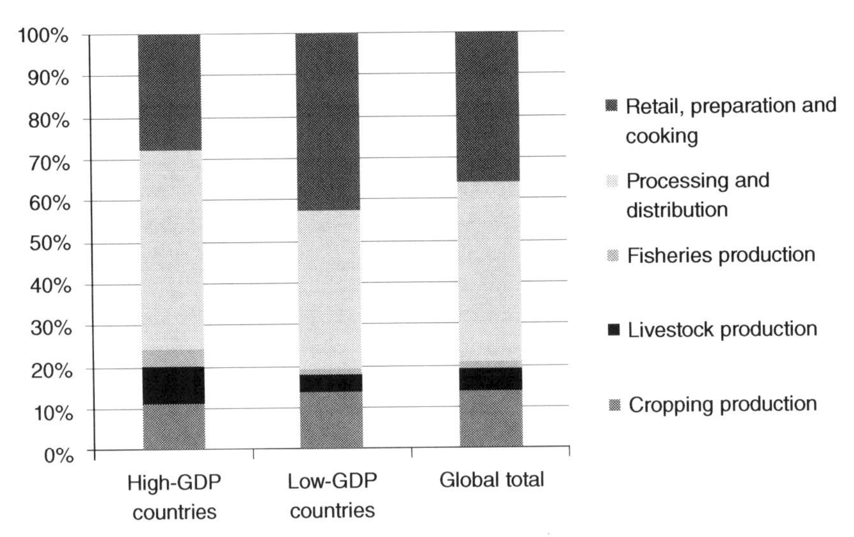

Figure 12.1 Direct and indirect energy inputs into the different segments of the food sector (by country classification) (FAO, 2011).

Renewable energy can play an important role in enhancing the resilience, reliability, price volatility insulation and efficiency of the supply chain. Combining food production with renewable energy generation is possible at the subsistence, small-scale and large-scale levels and can bring co-benefits to farmers, landowners, businesses and rural communities. For instance, millions of subsistence farmers use small-scale domestic digesters to produce biogas (FAO, 2011). Food processing plants can use biomass by-products for co-generating heat and power, which are usually consumed on-site. There is also potential to produce bioenergy from food processing plants.

Increased renewable energy utilization can reduce food wastage and embedded energy losses. In low-GDP countries most food losses occur during harvest and storage, as compared to high-GDP countries where food waste occurs mainly during the retail, preparation, cooking and consumption stage of the food supply chain. A significant share of total energy inputs are embedded in these losses. An age-old solution for reducing food losses and ensuring off-season availability is food preservation through drying. In comparison to other food preservation techniques, food drying stands out because it can be performed using low temperature thermal sources such as solar or geothermal energy, and the dried food produced can be easily stored and transported and has an extended shelf life (FAO, 2011).

Widely cited trade-offs between food security and bioenergy production can be resolved through greater regulations and technological advancement. Specific concerns have been raised on the sustainability of tapping into vast bioenergy resources, particularly within the transport sector. As such, the sustainability of biomass feedstock production, associated land-use changes, trade restrictions and the impacts of biofuels produced from food crops, such as corn, remain under review. For example, ethanol production in the United States consumes about 10 percent of annual global corn production, raising concerns about its impact on food supply. Potential trade-offs with food production need to be carefully considered and can be addressed through sustainable feedstock procurement practices and advances in biofuel production technologies.

In the transport sector, biofuels are being touted as a key alternative to fossil fuels. Liquid biofuels provide around 3.4 percent of fuels used in road transport and a very small amount of aviation fuels. Some economic entities such as Brazil, the United States and the European Union have higher shares, at 20.1 percent, 4.4 percent and 4.2 percent respectively (IEA, 2013). It is estimated that biofuel production will rise to 2.36 million barrels of oil equivalent per day by 2018, up from 1.86 million today, which would then account for 4 percent of total road transport fuel. Advanced biofuels, which use agricultural waste (with limited trade-offs with food production), had a global capacity at 4.5 billion litres by the end of 2012, which is a mere 77,500 barrels of oil equivalent per day.

The impact of renewable energy on land resources – a vital input to food production – is generally positive. Different renewable energy technologies impact land-use patterns differently. Solar and wind technologies are usually more land-intensive but have limited permanent impacts on the land. Moreover, their distributed nature means that they have a negligible land footprint (e.g. rooftop PV) and can support dual-usage of land. Other technologies, such as geothermal and ocean, are more site-specific in nature depending on resource availability. Hydro and biomass can have more significant impacts due to shifting land-use patterns. Hydropower plants with a water reservoir occupy large areas compared to run-of-river plants. Similarly, biomass power plants can have very diverse land-use requirements, depending on whether the feedstock is purpose-grown, in which case land use can be significant, or whether the feedstock is waste from forest or agriculture industries, in which case the only land required is the site for the power plants.

Adopting both a qualitative and quantitative approach to assessing the impact of renewables on land use is necessary. In the absence of either, any comparison remains incomplete and oftentimes biased. In assessing the land-use impacts of energy extraction and production, for instance, several renewable energy options, such as solar, wind and hydro, may appear to have a higher land footprint when seen only from a quantitative perspective. Their comparative benefits become more evident when qualitative metrics are also considered, such as the state of the land before and after (see Box 12.1).

Box 12.1 Renewable energy and land use

Land-use impacts can broadly be evaluated into three different categories: the area impacted, duration of impact and quality of impact (NREL 2009, 2013; Koellner and Scholz, 2008). A truly comprehensive assessment of land use requires an analysis that cuts across the three different categories.

Some studies conclude that onshore wind has the highest land use per unit of produced electricity (IEA, 2011). However, the turbines may occupy only 3–5 percent of the land and the rest can be employed for other uses, such as agriculture or grazing. When mining, processing and transport of coal are taken into account, the land use of coal-fired generation is comparable with that of solar technologies.

It is important to make a distinction between energy production techniques that clear all natural habitat within their area of impact. Coal and nuclear, for instance, cannot be placed in the same category as onshore wind or hydropower, given the scale of projects and the fact that onshore wind or hydropower have the potential for coexistence with productive farming or irrigation.

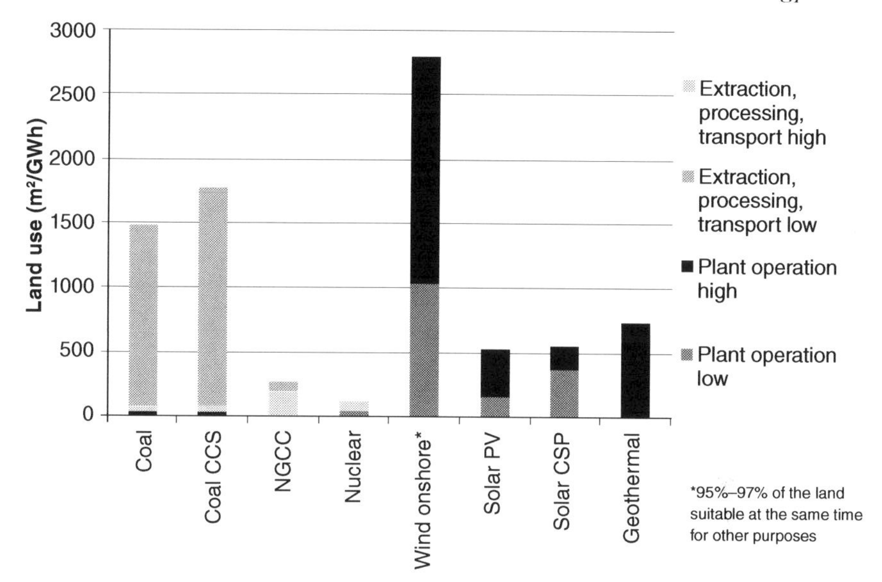

Figure 12.2 Land use requirements of power generation technologies (Fthenakis and Kim, 2010; MIT, 2006).

Renewable energy and the water–energy Nexus

The water and energy sectors stand to gain significantly from increasing renewable energy deployment. The existing stock of generation plants relies heavily on water as inputs for fuel extraction, processing and power production. Shifting weather patterns, changing climate and conflicting use of water resources are already posing challenges for the energy sector globally. Energy accounts for 15 percent of global water usage today and could consume 85 percent more through to 2035. Meeting growing demand for energy entirely through conventional fuels is no longer sustainable and requires alternatives that are less water-intensive with minimal impact on the environment.

Energy production accounts for about 15 percent of total water withdrawals, second only to agriculture. The vast majority of this water is used for cooling purposes during energy generation, creating potential constraints for traditional power generation stations. The fuel transformation supply chain requires water inputs at nearly all other stages as well (e.g. fossil fuel extraction, biomass feedstock growth, etc.) (IEA, 2012).

Many renewable energy technologies have significantly lower water intensity when compared to different energy production technologies. Figure 12.3 illustrates the water consumption factors for different electricity generation technologies.[2] Solar PV and wind use minimal amounts of water in the operations stage (for cleaning and maintenance) compared to conventional technologies, such as coal, nuclear or natural gas. This makes them well suited for a carbon- and water-constrained environment.

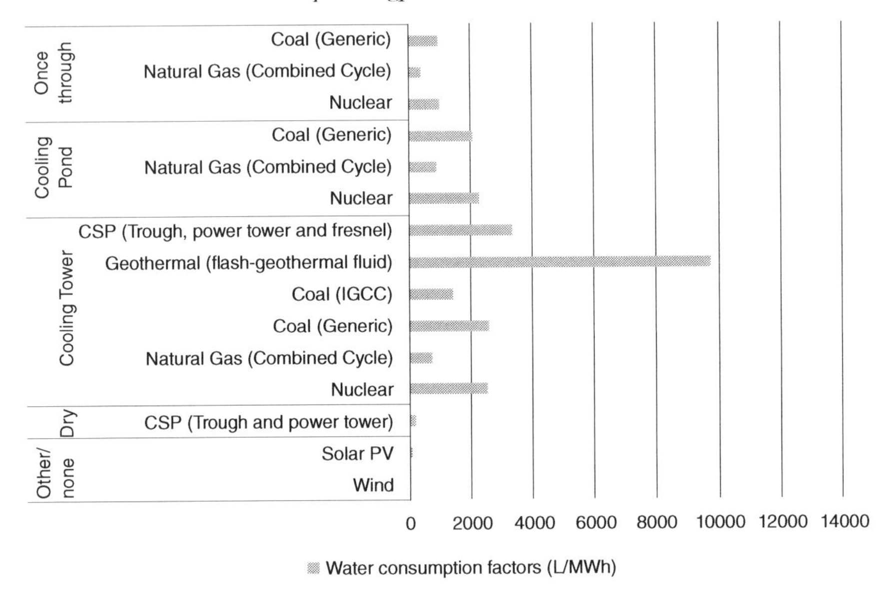

Figure 12.3 Water consumption factors for electricity generation technologies (NREL, 2011).

As lower water-intensity renewables expand, the cumulative water–energy decoupling impact will become even more significant. During 2013 electricity from wind energy in the United States has avoided the consumption of more than 130 billion litres of water, equivalent to the annual water consumption of over 320,000 households (AWEA, 2013).[3] A scenario with 20 percent wind energy in the US energy mix could reduce cumulative water use in the electricity sector by nearly 8 percent (US DoE, 2008). In China, where the coal industry accounts for over 22 percent of the total water consumption, massive renewables deployment is expected to reduce the reliance of power generation on water from 97 percent to under 87 percent by 2030 (Cho, 2011; Tan, 2012).

Geothermal and concentrated solar power (CSP) utilize thermoelectric generation and requires relatively more water during operation. For geothermal energy, water use estimates vary widely depending on technology. When geothermal fluids are included in water needs for operational water requirements, estimates for water consumption range from 4,000–16,000 litres per MWh and higher in some cases (Davies *et al.*, 2013; Meldrum *et al.*, 2013). Concentrated solar power is more water-intensive during the operation, in particular where steam turbines are prime movers. According to Burkhardt *et al.* (2011), CSP with wet cooling consumes more water (up to 4,700 L/MWh) than most thermoelectric power plants, although the application of dry-cooling systems can reduce the total water consumption by 77 percent during a plant's life cycle. Another source puts water withdrawal rates at around 3,030 L/MWh (Mielke *et al.*, 2010).

In the process of hydropower generation, water losses occur due to evaporation from holding reservoirs. Once evaporation is accounted for, large hydropower can be very water-intensive, consuming up to 17,000 litres to produce one MWh of electricity (Mielke *et al.*, 2010). However, reservoirs are used for many purposes, so allocating the entire losses to electricity generation may not be accurate. Small hydropower and run-of-the-river technologies do not need large dams, avoiding the large amounts of water evaporation and aquatic ecosystem impact associated with large hydropower.

While most renewable energy resources are harnessed for end use with minimal conversion processes, others, such as bioenergy, necessitate feedstock growth, processing and transport, often requiring substantial water inputs. Water inputs vary largely by crop type, local climatic conditions and technology choices. Irrigated feedstock requires large volumes of water, while agricultural and forestry residues do not require high additional water or land. On average 4,500 litres of water is consumed per MMBtu of ethanol produced for the feedstock growth stage (Water in the West, 2013). However, considerable improvements can be made in water use efficiency in bioenergy crops by improving water retention and lowering direct evaporation from soils. Ethanol refining averages 159 litres per MMBtu of ethanol, and this amount is falling rapidly: dry mills consume 50 percent less water during refining than a decade ago (Wu *et al.*, 2009; RFA, 2013). Emerging advanced biofuels, made from lignocellulosic biomass or woody crops, agricultural residues or waste, are expected to have lower water input requirements and reduced competition with food crops because they mainly use perennial plants as feedstock and utilize crop residues currently not used (Mielke *et al.*, 2010).

Renewables can enhance water security by improving the reliability and sustainability of water systems, and reducing contamination threats.[4] Water use is inextricably linked to energy supply, requiring energy inputs for every step. Figure 12.4 depicts this supply chain and identifies the various segments where renewable energy can replace conventional energy sources, thus easing the strains on the water and energy Nexus. Renewable energy technologies also offer decentralized solutions to improve access to water supply, especially in rural and remote areas.

Water pumping forms a critical part of the supply chain to extract groundwater, an essential source of life and livelihood for the world's population. Water pumping can be energy-intensive, requiring large amounts of electricity or diesel fuel. In India, for instance, the world's largest user of groundwater for agriculture, an estimated 87 billion kWh/year, or 15–20 percent of total national energy, is used by electric pumps. Another 4–4.5 billion litres of diesel fuel is used annually in unelectrified areas for irrigation purposes (Shah, 2008).[5]

Increasing reliance on diesel and subsidized electricity for water pumping can negatively impact water and energy security. Fluctuating diesel prices can impact the affordability and reliability of pumping services. For instance, high diesel prices forced the shutdown of several pumps on the Ogallala aquifer in

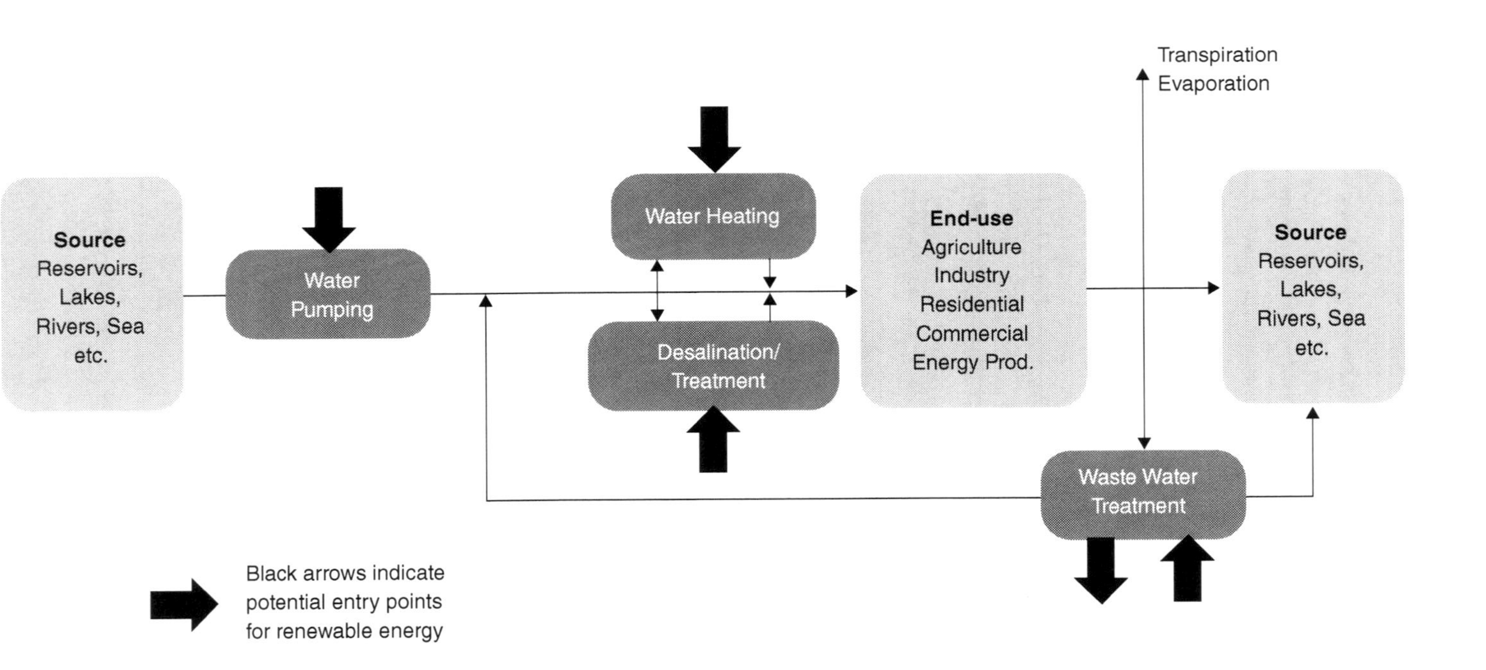

Figure 12.4 Potential areas for renewable energy intervention across the water supply chain.

the US when diesel prices were three times as high as electricity prices (Zhu *et al.*, 2007). In the South Asia region as a whole, regional energy providers spend a cumulative US$3.78 billion to subsidize groundwater pumping for irrigation. This subsidized energy has also been a major factor in overdrawing of non-renewable groundwater sources. As water levels drop, more power is needed to retrieve water, thus increasing stresses on the energy system.

Solar PV pumping systems are beginning to be widely used in the developing world due to their increasing competitiveness compared to diesel and grid-connected electrical pumps. For example, Tunisia's 2008 Renewable Energy plan included measures to develop renewable energy in the agricultural sector, with two hundred large water pumping stations for irrigation systems powered by hybrid technologies (REN21, 2013). Similarly, India has committed to replace 26 million of its groundwater pumps with more efficient, solar powered pumps, which is estimated to replace US$6 billion in combined electricity and diesel subsidies annually (Pearson and Nagarajan, 2014). While solar pumping reduces reliance on fossil fuels, it needs to be accompanied by regulatory measures to minimize overdrawing of water resources.

In many parts of the world where renewable water resources are limited, increased reliance on energy-intensive water supply options such as desalination is expected (World Bank, 2012). Desalination is generally energy-intensive and reliant on fossil fuels, which are vulnerable to volatile energy markets and logistical challenges in remote areas. At present over 14,000 desalination plants operate globally, providing several billion gallons daily (Shatat, Worall and Riffat, 2013). In Qatar, for example, 99 percent of water is procured by desalination. Desalination capacity is expected to grow by 9 percent annually, with most growth in the Gulf Cooperation Council countries, Asia and the Caribbean. The Gulf region relies primarily on thermal desalination, often the most energy-intensive technology for water procurement. This means that many countries spend as much as a quarter of their energy produced on water production through desalination.

Renewable energy technologies can support desalination. Concentrated Solar Power (CSP) technology is most compatible with thermal desalination, requiring high heat inputs. Membrane desalination utilizes electric motors and can be powered by solar PV, CSP, wind and geothermal energy. Currently, only 1 percent of total desalinated water is produced by energy from renewable sources (IEA-ETSAP and IRENA, 2012). Saudi Arabia has recently announced targets to make all desalination run from solar power by 2020. Small-scale desalination plants are already proving cost-effective in remote regions where the cost of delivery is high, as illustrated in Table 12.1 (Shatat, Worall and Riffat, 2013).

Table 12.1 Costs for comparative desalination technologies (Shatat, Worall and Riffat, 2013).

Type of feed water	*Type of energy*	*Water cost (US$/m³)*
Brackish water	Conventional fuel	0.27–1.38
	PV	5.85–13.42
	Geothermal	2.6
Sea water	Conventional fuel	0.46–3.5
	Wind energy	1.3–6.5
	PV	4.08–11.7
	Solar collectors	4.55–10.40

Challenges for renewable energy in the Nexus

Renewable energy offers significant comparative advantages over conventional alternatives in managing the Nexus. The potential exists but challenges remain:

1. There is a general lack of scientific data or national-level statistics to conduct a system-level assessment of the benefits (and challenges) renewables present from a Nexus perspective.
2. A proper assessment of options requires taking into account the life cycle perspective of energy technologies. As our chapter describes, the scientific data required for such assessment still requires development and peer review.
3. The global energy landscape is currently undergoing a fundamental transformation with profound impacts on incumbent players, not all of the effects positive. With increasing complexity, a higher degree of decentralization and many new players in the market, finding Nexus-friendly pathways for growth may become more challenging. The challenge is to find pathways for growth that are inclusive and provide win–win situations for most.
4. Some renewable energy technologies require water or land and potentially compete with food under certain circumstances. Technologies for cost-competitive advanced biofuels need to be developed, the cost for dry cooling for CSP plants needs to be reduced and dual uses of dam-based hydro or solar/wind energy farms promoted.

The dynamics driving the transformation of the energy sector, such as increased demand, environmental concerns and threats to security of supply, apply in most parts also to the water and agriculture sector. With adequate planning and coordination, renewable energy technologies can be a collective solution for the three sectors. Realizing that potential, however, will require rethinking across the different levels in terms of how resources have traditionally been allocated, distributed and consumed.

Notes

1 The Food and Agriculture Organization (FAO) defines food security as existing "when all people, at all times, have physical, social and economic access to sufficient, safe and nutritious food that meets their dietary needs and food preferences for an active and healthy life" (FAO, 2002).
2 Water withdrawal is defined as "water diverted or withdrawn from a surface water or groundwater source." Water consumption is the "water use that permanently withdraws water from its source; water that is no longer available because it has evaporated, been transpired by plants, incorporated into products or crops, consumed by people or livestock, or otherwise removed from the immediate water environment" (Vickers, 2001).
3 Assumes average household consumption of 1,135 litres per day (US EPA, n.d.).
4 Water Security is defined as "The capacity of a population to safeguard sustainable access to adequate quantities of and acceptable quality water for sustaining livelihoods, human well-being, and socio-economic development, for ensuring protection against water-borne pollution and water-related disasters, and for preserving ecosystems in a climate of peace and political stability" (UN Water, 2013).
5 In the United States, the next biggest user of groundwater for agriculture, groundwater pumping alone constitutes 1–2 percent of national electricity consumption (Water in the West, 2013).

References

AWEA (American Wind Energy Association) (2013) *Wind Energy Conserving Water*. Available online at: www.awea.org/windandwater

Burkhardt, J.J., Heath, G.A. and Turchi, C.S. (2011) Life cycle assessment of a parabolic trough concentrating solar power plant and the impacts of key design alternatives. *Environmental Science & Technology* 45: 2457-2464

Cho, R. (2011) *How China is Dealing With Its Water Crisis*. Available online at: http://blogs.ei.columbia.edu/2011/05/05/how-china-is-dealing-with-its-water-crisis

Davies, R.G.E., Kyle, P. and Edmonds, A.J. (2013) An integrated assessment of global and regional water demands for electricity generation to 2095. *Advances in Water Resources*, available online at: www.researchgate.net/publication/240917536_An_integrated_assessment_of_global_and_regional_water_demands_for_electricity_generation_to_2095

EU ERC (European Union European Research Council) (2011) *Causes of the 2007–2008 global food crisis identified*. Available online at: http://ec.europa.eu/environment/integration/research/newsalert/pdf/225na1.pdf

FAO (Food and Agriculture Organisation) (2002) Chapter 2. Food security: concepts and measurement. Available online at: www.fao.org/docrep/005/y4671e/y4671e06.htm

FAO (2011), *Energy-Smart Food for People and Climate*. Rome: FAO. Available online at: www.fao.org/docrep/014/i2454e/i2454e00.pdf

Fthenakis, V. and Kim, H. (2010) Life-cycle uses of water in US electricity generation, *Renewable and Sustinable Energy Reviews*, 14(7), 2039–2048. Available online at: www.sciencedirect.com/science/article/pii/S1364-0321(10)00063-8

IEA (International Energy Agency) (2011) Renewable energy: Policy considerations for deploying renewables. Available online at: www.iea.org/publications/freepublications/publication/Renew_Policies.pdf

IEA (2012) *World Energy Outlook 2012*, Paris: IEA.

IEA (2013) *World Energy Outlook 2013*, Paris: IEA.

IEA-ETSAP and IRENA (2012), Water desalination using renewable energy: Technology brief. Available online at: www.irena.org/DocumentDownloads/Publications/IRENA-ETSAP percent20Tech percent20Brief percent20I12 percent20Water-Desalination.pdf

Koellner, T. and Scholz, R. (2008) Assessment of land use impacts on the natural environment. *The International Journal of Life Cycle Assessment*, 1(1), 32–48.

Meldrum, J., Nettles-Anderson, S., Heath, J. and Macknick, J. (2013) Life cycle water use for electricity generation: A review and harmonization of literature estimates. Available online at: http://iopscience.iop.org/1748-9326/8/1/015031/article

Mielke, E., Anadon, L.D. and Narayanamurti, V. (2010) Water consumption of energy resource extraction, processing and conversion. Available online at: http://belfercenter.ksg.harvard.edu/files/ETIP-DP-2010-15-final-4.pdf

MIT (2006) The Future of Geothermal Energy. Available online at: http://geothermal.inel.gov/publications/future_of_geothermal_energy.pdf

NREL (2009) *Land-use Requirements of Modern Wind Power Plants in the United States*, Golden: NREL.

NREL (2011) A review of operational water consumption and withdrawal factors for electricity generating technologies. Available online at: www.nrel.gov/docs/fy11osti/50900.pdf

NREL (2013) *Land-use requirements for solar power plants in the United States*. Available online at: www.nrel.gov/docs/fy13osti/56290.pdf

OECD (2012) *OECD Environmental Outlook to 2050*, OECD Publishing. Available online at: http://dx.doi.org/10.1787/9789264122246-en

Pearson, O. and Nagarajan, G. (2014) *Solar Water Pumps Wean Farmers From India's Archaic Grid*, New York: Bloomberg. Available online at: www.bloomberg.com/news/articles/2014-02-07/solar-water-pumps-wean-farmers-from-india-s-archaic-grid

REN21 (2013) *Renewables 2013: Global Status Report, Renewable Energy Policy Network for the 21st Century*. Available online at: www.ren21.net/Portals/0/documents/Resources/GSR/2013/GSR2013_lowres.pdf

Renewable Fuels Association (RFA) (2013) *Ethanol and the US Corn Crop*. Available online at: www.ethanolrfa.org/page/-/objects/documents/1898/corn_use_facts

Shah, T. (2008) Climate change and groundwater: India's opportunities for mitigation and adaptation. Available online at: http://iopscience.iop.org/1748-9326/4/3/035005/pdf/erl9_3_035005.pdf

Shatat, M., Worall, M. and Riffat, S. (2013) Opportunities for solar water desalination worldwide: Review, *Sustainable Cities and Society*, 9, 67–80. Available online at: http://dx.doi.org/10.1016.j.scs.2013.03.004

Tan, D. (2012) *China: No Water, No Power*. Available online at: http://chinawaterrisk.org/resources/analysis-reviews/china-no-water-no-power/

UN Water (2013) *Water Security and the Global Water Agenda*, Geneva: UN.

US DoE (United States Department of Energy) (2008) *20 percent Wind Energy by 2030: Increasing Wind Energy's Contribution to US Electricity Supply*. Available online at: *www.nrel.gov/docs/fy08osti/41869.pdf*

US EPA (United States Environmental Protection Agency) (n.d.) Available online at: www.epa.gov/watersense/our_water/water_use_today.html

Vickers, A. (2001) *Handbook of Water Use and Conservation: Homes, Landscapes, Businesses, Farms*, Amherst, PA: WaterPlow Press.

Water in the West (2013) *Water and Energy Nexus: A Literature Review*. Available online at: http://waterinthewest.stanford.edu/sites/default/files/Water-Energy_Lit_Review.pdf

World Bank (2012) *Renewable Energy Desalination: An Emerging Solution to Close the Water Gap in the Middle East and North Africa*. Washington, DC: World Bank. DOI: 10.1596/978-0-8213-8838-9. License: Creative Commons Attribution CC BY 3.0.

World Bank (2013) *Thirsty Energy: Securing Energy in a Water-Constrained World*. Washington: World Bank. Available online at: www.worldbank.org/en/topic/sustainabledevelopment/brief/water-energy-nexus

Wu, M., Mintz, M., Wang, M. and Arora, S. (2009) Water consumption in the production of ethanol and petroleum gasoline, *Environmental Management*, 44(5), 981–997.

WWAP (United Nations World Water Assessment Programme) (2014) *The United Nations World Water Development Report 2014: Water and Energy*. Paris: UNESCO.

Zhu, T., Ringler, C. and Cai, X. (2007) Energy price and groundwater extraction for agriculture: Exploring the Energy Water Food Nexus at the global and basin levels, *International Water Management Institute*. Available online at: www.iwmi.cgiar.org/EWMA/files/papers/Energyprice_GW.pdf

13 Adding to complexity

Climate change in the Energy–Water Nexus

Diego Rodriguez, Anna Delgado Martin and Antonia Sohns

Background

According to the US National Oceanic and Atmospheric Administration (NOAA), 2014 was the hottest year on record. For the year to date, global temperatures have measured 1.22 degrees Fahrenheit above the twentieth century average of 57.3 degrees Fahrenheit, making the period from January to August 2014 the third warmest period since records began in 1880 (Thompson and Climate Central, 2014). The 2013 report from the Intergovernmental Panel on Climate Change (IPCC) emphasized that the planet's climate has been unequivocally warming since the 1950s, with many of the observed changes unprecedented over decades to millennia. "The atmosphere and ocean have warmed, the amounts of snow and ice have diminished, sea level has risen, and the concentrations of greenhouse gases have increased" (IPCC, 2013). If measures are not taken and countries continue to emit greenhouse gases (GHGs) at sustained levels, the systems societies and ecosystems depend on will be threatened by increased temperatures, and the impacts of climate change will have dramatic consequences for food, water and energy systems.

Climate change heightens the likelihood of extreme weather conditions such as prolonged drought periods and major floods. The contrast between wet and dry regions and wet and dry seasons will also increase as a result of climate change, enhancing water, energy and food insecurity and placing populations, livelihoods and assets in danger (IPCC, 2013). The effects and intensity of climate change will vary regionally as populations experience a change in average precipitation, surface runoff and stream flow, deviation from rainfall averages, and increased probability of extreme events such as intense storms, floods and droughts. Altered precipitation and evapotranspiration patterns are predicted to reduce runoff in southern Africa, the Mediterranean basin, Central America, the southwestern United States and Australia, among other places (FAO, 2008). This is likely to increase competition for water across sectors such as agriculture, energy, water supply and the environment.

Climate change will result in a heightened reliance on energy-intensive water supply options such as water transfers, increased water pumping, or desalination plants to supplement urban water supply. As temperatures rise more water will be needed by the energy sector to meet both its own demand for water for cooling per unit of energy produced, and also to meet increased energy demands for the cooling of houses, offices and factories. On the other hand, the energy sector accounts for 40 percent of the global CO_2 emissions every year (IEA, 2012a), contributing to the same climate change that is impacting the sector. The energy sector needs to be part of the solution and the transition to energy supply options involving low or zero greenhouse gas emissions. However, some of these options, such as carbon capture and storage or irrigated biofuels, could further increase pressure on water resources (Newmark *et al.*, 2010; WWAP, 2014; Wallis *et al.* 2014).

Today, about 748 million people still lack access to improved sources of drinking water – nearly half of these are in sub-Saharan Africa – and more than one-third of the global population, around 2.5 billion people, remain without access to improved sanitation (WHO/UNICEF, 2014). Additionally, over 1.3 billion people lack access to electricity, with most of them residing in sub-Saharan Africa and South and East Asia (IEA, 2012a). In the coming decades, these statistics are predicted to worsen. It is expected that by 2025, 1.8 billion people will live in countries or regions with absolute water scarcity and two-thirds of the world population could be affected by water stress conditions (WWAP, 2012). Global energy consumption will increase by nearly 35 percent by 2035 (IEA, 2012b), which can also impact already-strained water resources. Population and economic growth will place additional burdens on food, energy, and water resources (WWAP 2014).

The energy sector already recognizes the magnitude of climate change's impacts on their operations (Rubbelke and Vogele, 2011; van Vliet *et al.*, 2012; US DOE, 2013). In particular, the relationship between water and energy is one of shared tensions and dependencies. Significant amounts of water are needed in almost all energy generation processes, from generating hydropower to cooling and other purposes in thermal power plants, to extracting and processing fuels (WEC, 2010; Stillwell *et al.*, 2011). Conversely, the water sector needs energy to extract, treat and transport water (Goldstein and Smith, 2002; ESMAP, 2012). Both energy and water are used in the production of crops, including those used to generate energy through biofuels (FAO, 2012). Energy and water use by industry, agriculture and for domestic consumption can have adverse impacts on ecosystems through the loss of habitat and changes in biological processes (WBCSD, 2009). This relationship is what is known as the Water–Energy Nexus (Pate *et al.*, 2007; Webber, 2008; SEI, 2011). These interdependencies complicate possible solutions and make a compelling case to expeditiously improve integrated water and energy planning in order to avoid unwanted future scenarios (Rodriguez *et al.*, 2013; WWAP, 2014).

Climate change exacerbates the underlying linkages between water and energy, adding uncertainty and risks that, if not addressed now, will compromise development and push people further into poverty. For example, rain patterns will shift and intensify, thereby enhancing uncertainty in energy development.

Increasing global temperatures will warm rivers and lakes that power plants rely on for cooling water, thus making electricity generation harder in the future. In some cases future water scarcity will threaten the viability of projects and hinder development (IEA, 2012b).

Existing and future Water–Energy–Climate challenges

Although the Water–Energy (and Climate) Nexus varies by region, challenges in securing enough water for energy and energy for water will increase with population and economic growth. The relationship between water and energy is already critical in many regions, and the resulting stresses are compounded as demand grows from emerging economies and developing countries. Water constraints have already adversely impacted the energy sector in many parts of the world. In the US several power plants have been affected by low water flows or high water temperatures (US DOE, 2010; US DOE, 2013). In India a thermal power plant recently had to shut down due to a severe water shortage (*Times of India*, 2014). France has been forced to reduce or halt energy production in nuclear power plants due to high water temperatures threatening cooling processes during heatwaves (World Economic Forum, 2011). Thermal power plant projects and fuel extraction facilities are being re-examined due to their impact on regional water resources and their vulnerability to climate impacts (UNEP, 2010; UNEP, 2012a).

Projected consequences of these factors are alarming enough to require the urgent development of more accurate integrated planning tools. In 2012 the International Energy Agency's World Energy Outlook report included a special section on the water needs and the possible future water constraints of the energy sector. The report concluded: "constraints on water can challenge the reliability of existing operations and the viability of proposed projects, imposing additional costs for necessary adaptive measures" (IEA, 2012b). The CDP (Carbon Disclosure Project) Global Water Report 2013 highlighted that 82 percent of energy companies and 73 percent of power utility companies indicate that water is a substantive risk to business operations; and that 59 percent of energy companies and 67 percent of power utility companies have experienced water-related business impacts in the past five years (CDP, 2013). Yet the CDP report found that most companies do not have a water strategy; instead they opt for a "business-as-usual" approach. A UN Environment Program Water Report surveyed more than 125 countries on this topic and found that the issue of water for energy was high or very high on the list of priorities in 48 percent of the countries surveyed (UNEP, 2012b). A recent report by the US Department of Energy (US DOE, 2013) highlighted energy sector vulnerabilities to climate change and extreme weather, finding that most risks were water-related.

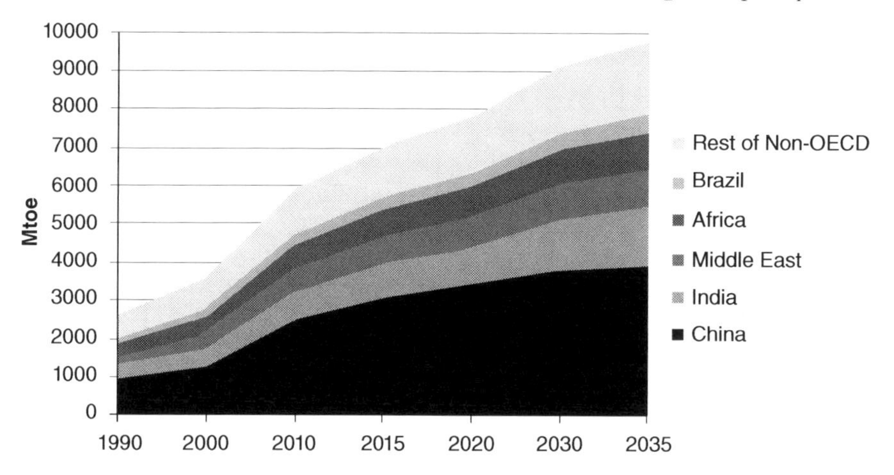

Figure 13.1 Non-OECD primary energy demand by region. Mtoe stands for million tons of oil equivalent (based on IEA, 2012b).

Developing countries are the most vulnerable

The vast majority of future population growth will be in developing countries (UN, 2010). This, together with economic growth and rapid urbanization, will place huge demands on both energy and water resources (see Figure 13.1).

Moreover, developing countries are the most vulnerable to the consequences of increased demand and of climate change because they often lack infrastructure to mitigate increasingly variable water supply (World Bank, 2012a). Developing countries often also lack the institutions and regulatory framework to manage this complexity.

While the challenge is clear, current energy planners and governments overlook these concerns and often make planning decisions without taking into account existing and future water constraints. Planners in both sectors often remain ill-informed about the drivers of these challenges, how to address them, and the merits of different technical, political, management and governance options. The absence of integrated planning between these two sectors is socio-economically unsustainable. Integrated water and energy planning will become crucial to ensure a sustainable future strategy for many countries, especially in areas of the world where climate change is going to exacerbate water scarcity.

Why does the energy sector need water?

Most thermal power plants require large amounts of water for cooling. Thermal power plants generate around 80 percent of the electricity in the world (IEA World Energy Outlook 2013) by converting heat into power in the form of electricity.[1] Most of them heat water to transform it into steam, which then spins the turbines that produce electricity (Wolfe *et al.*, 2009). After passing through the turbine the steam is cooled down and condensed to start the cycle again (closing the steam cycle) (see Figure 13.2). The water is heated with different energy sources (coal, oil, natural gas, uranium, solar energy, biomass, geothermal energy) depending on the sub-type of power plant, but the principle is the same. All these power plants need to cool down the steam, and most of these power plants use water to do so (through wet cooling systems), which requires them to be near a water source (a river, lake or ocean).

Thus, although thermal power plants require water for several processes (such as the steam cycle, ash handling and flue gas desulfurization systems), the largest quantities of water are required for cooling purposes (DiPietro *et al.*, 2009; US DOE, 2010; Macknick *et al.*, 2011).[2] There are different types of cooling systems: once through, closed loop, dry cooling or hybrid. *Once through cooling* is the simplest and cheapest type of cooling system. It requires large amounts of water, but consumes a very small fraction of it. The water is withdrawn from the water source and run once through the power plant (hence the name) to cool down the steam. The water is then discharged back into the source a few degrees warmer, which can cause thermal pollution. Since such systems withdraw large quantities of water they can also kill fish and other organisms in the process. Moreover, this is the type of cooling system that is affected the most during drought or very hot seasons (van Vliet *et al.*, 2012; Stewart *et al.*, 2013). *Closed loop cooling systems* withdraw much less water but consume most of it (the water is evaporated). These cooling systems are more complex and expensive than once-through systems, but since they withdraw less water they have less environmental impact and are less susceptible to drought. *Dry cooling systems* use air instead of water to cool down the steam. Thus, they do not consume any water. However, their high cost and the negative impact on power plant efficiency are still drawbacks for this type of system, which is rarely used. There are also *hybrid cooling systems* which combine wet and dry cooling approaches (Maulbetsch, 2011). The cooling system employed by a power plant has an impact on plant efficiency, capital and operation costs, water consumption, water withdrawal, and total environmental impacts. Therefore, trade-offs must be evaluated on a case-by-case basis, taking into consideration regional and ambient conditions and existing regulations.

Therefore, the amount of water withdrawn and consumed by the power plant will mainly depend on the type of cooling system used. Other factors such as climate also matter, but they have a smaller impact on the amount of water used. Among plants with the same type of cooling systems, the amount of cooling water withdrawn and consumed is mainly determined by the power plant's efficiency (Delgado Martin, 2012). More efficient power plants require less water than older and less efficient plants. Given that not all power plants grouped under a type (for example, "coal subcritical power plant") have the same efficiency and neither are they located in areas with the same climate, there is no single "water factor" (i.e. water requirement per unit of energy produced) for each specific type of power plant (Macknick *et al.*, 2011). The

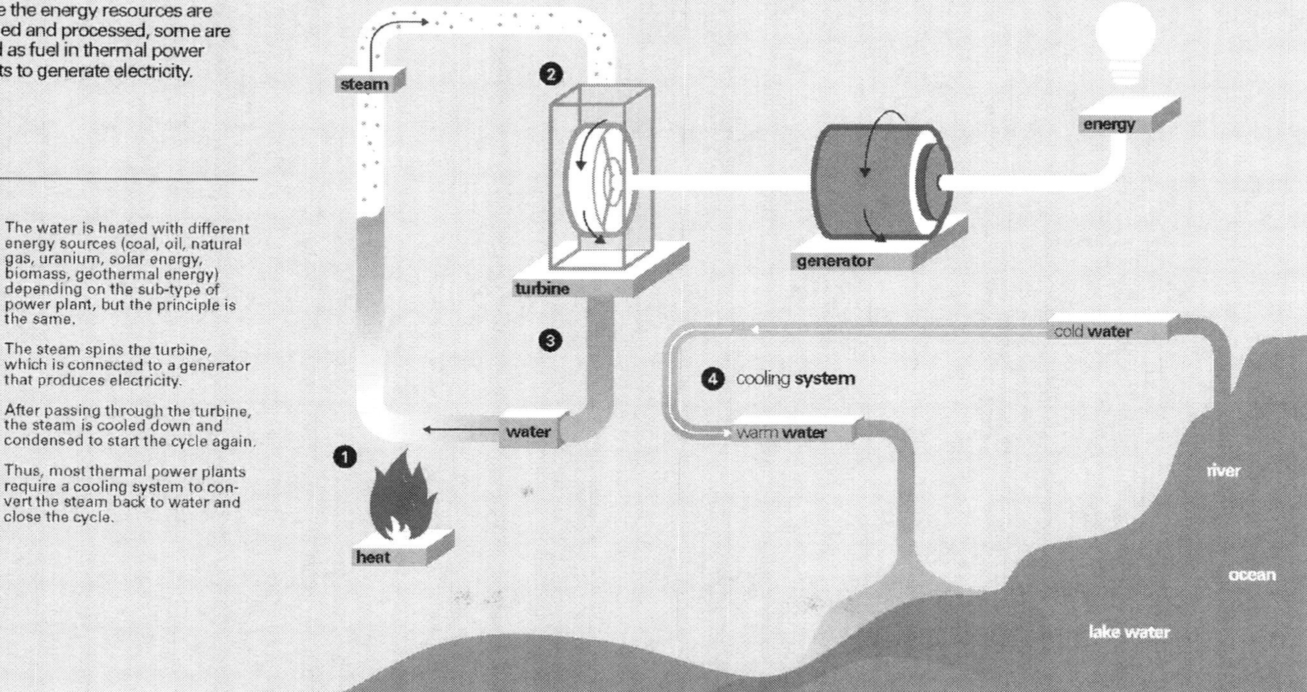

Figure 13.2 Diagram of the steam cycle (Thirsty Energy, World Bank: www.worldbank.org/thirstyenergy).

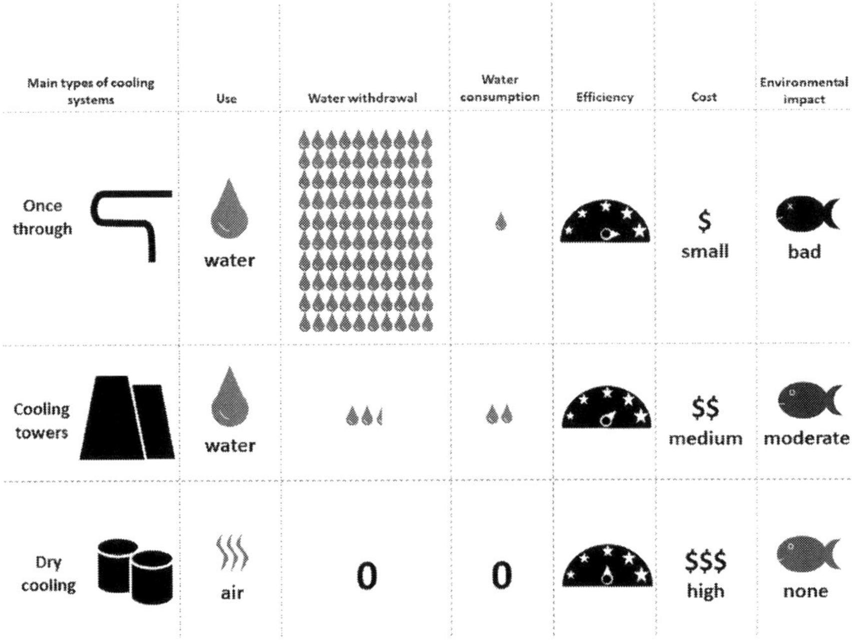

Main types of cooling systems	Use	Water withdrawal	Water consumption	Efficiency	Cost	Environmental impact
Once through	water				$ small	bad
Cooling towers	water				$$ medium	moderate
Dry cooling	air	0	0		$$$ high	none

Figure 13.3 Trade-offs between different types of cooling systems (Thirsty Energy, World Bank: www.worldbank.org/thirstyenergy).

vast differences in water demand in the energy sector impose an important challenge when analyzing and quantifying potential constraints.

Hydropower uses the energy of moving water to turn the turbines and generate electricity. Thus, hydropower planners have been aware of the "Water–Energy Nexus" for a long time, since they depend on the availability of water resources to generate electricity. In hydropower plants, most of the water is not consumed but only diverted to generate electricity. Loss of water is normally small and is only caused by evaporation from the dammed water upstream of the plant. Thus, most of the water can be used downstream for other purposes, such as irrigation and urban water supply. Run-of-river hydropower plants, which store no water, have water evaporative losses near zero. However, a run-of-river site cannot be used for generation of peak loads or during dry seasons when there is no or little river flow. Hydropower dams, which can provide power throughout the year and balance peak demand, can consume some water through evaporative losses from the reservoir but this consumption varies greatly depending on site location, design and operation (WWAP, 2012). Despite the losses from reservoirs, however, hydropower dams may increase water availability for downstream users when it is needed the most, e.g. in a drought situation. In a world of severe energy shortages and increasing water variability, hydropower dams and their multipurpose water infrastructure will play an expanding role in providing electricity and allocating scarce water resources. The joint equitable use and careful planning of sustainable power and water infrastructure in river basins is key for the Water–Energy Nexus challenge.

Wind turbines do not require any water for their operation. Photovoltaic (PV) solar systems require minimal quantities of water for mirror washing, which can nonetheless be challenging since these systems are usually located in sunny and arid places. Moreover, these power generation technologies face another problem: intermittency. Absent massive grid-scale electricity storage, it seems inevitable that there will be a need to continue using thermal power plants and hydropower as base-load and dispatchable power.

In the case of energy extraction, water is required for acquiring the resource (such as oil, gas, coal, biomass), processing and refining, and transportation. Energy development can greatly impact water resources through water discharged post-production or via possible contamination of aquifers during drilling, among other concerns (USGS, 2013). Thus, water used during extraction and wastewater generated from production must be managed carefully in order to ensure protection of the environment and the resources in the long term.

Water is also increasingly used for the production of biofuels. The impact of biofuels and biodiesel on water use varies greatly, depending on where the biofuel is planted, whether it requires irrigation, and if it replaces a more or less water-intensive crop and therefore affects local water demand (FAO, 2008).

Water risks for the energy sector

There are many water- and climate-related risks that can affect the energy sector and may slow or hinder the progress made towards energy goals. Increased water temperatures are correlated with rising air temperatures (Stewart *et al.*, 2013) and can prevent power plants from cooling properly, causing them to shut down or decrease production and incurring financial and economic losses (van Vliet *et al.*, 2013). Water scarcity, variability and quality may constrain or drive up the operational costs of power generation (affecting thermal power plants and hydropower), as well as affecting fuel extraction processes due to their large water requirements, although in most cases the cost of accessing water is very small in comparison with the revenue generated. Regulatory uncertainty is also a risk for the energy sector (CDP, 2013), including restricted operational water permits, rising discharge compliance costs and withdrawal limits. Moreover, sea level rise could adversely impact coastal energy infrastructure and power plant operations, and climate change will also affect the energy sector through varied energy demand, especially for cooling homes, offices and factories as temperatures increase.

Van Vliet *et al.* have analyzed the climate change impacts on the power sector in Europe and the US, where thermal power plants account for 78 percent and 91 percent of the electricity produced respectively, and 43 percent and 41 percent of the total freshwater withdrawals (Kenny *et al.*, 2009; van Vliet *et al.*, 2012). They found that the thermoelectric power sector is vulnerable to climate change (due to the combined impacts of lower summer river flows and higher river water temperatures) and concluded that on average, during the summer, power plants' capacity could decrease by 6.3 to 19 percent in Europe and by 4.4 to 16 percent in the United States, depending on the cooling system type and climate scenario for 2031–2060 (van Vliet *et al.*, 2012). They also projected higher wholesale electricity prices for most European countries as these limitations in water availability affect power plants (van Vliet *et al.*, 2013). Research like this will become increasingly important as companies consider alternative technologies (such as dry cooling), and governments study the placement of power plants along rivers, ensuring the plants' sustainable future operation under increased energy demand and in a potentially warmer climate.

There are many alternatives to address the Water–Energy Nexus in power generation, such as better cooling system technologies. However, many current options are less efficient and more costly, so operators prefer conventional systems until regulation or pricing dictate otherwise. However, if future climate risks (and the costs that these risks might cause to the energy sector) were taken into account in the decision making process today, we would avoid unwanted scenarios in the future. Due to these risks, governments must re-examine where power plant projects and energy extraction facilities are located, which types of energy generation technologies are chosen and which cooling systems are employed, to avoid locking into an infrastructure that might not be suitable and flexible enough for the changing future.

Trade-offs and regional differences

Unlike the emission of greenhouse gases, which is intrinsically a global problem, the Water–Energy Nexus is a regional and local problem. The regional variations in water resources, population and economic growth, water demand, electricity demand, energy mix and climate change all combine to create "hot-spots" where the Water–Energy Nexus is more complex than elsewhere. It is important to understand the regional challenges and to specifically address the Nexus in those critical regions.

As such, a "one solution fits all" approach is not appropriate for water–energy challenges. Solutions need to be contextualized and localized for enduring success. It is only through data collection coupled with sound analysis that a deeper understanding of systems that affect water (hydrological, socio-economic, financial, institutional and political alike) can be factored into water governance. It is also crucial to understand the complex trade-offs and implications for other sectors, in order to properly understand unintended consequences of certain energy and water policies.

GHGs vs water: Policies to reduce GHG emissions might impact water resources (Wallis *et al.*, 2014) and vice versa. In power plants, dry-cooling systems require no water for their operation, but are more costly than conventional cooling systems and decrease the efficiency of the plant, thus increasing GHG emissions per kWh (Maulbetsch and DiFilippo, 2006), further impacting climate change. CCS (Carbon Capture and Storage) technologies can reduce CO_2 emissions in the power sector, but since CCS reduces the efficiency of the power plant and requires cooling for its own processes, more water is required to achieve the same energy output (Newmark *et al.*, 2010; EPRI, 2010; Chandel *et al.*, 2011). Biofuels might be an effective way of reducing GHGs emitted by the transport sector, but if the biofuels are irrigated this may add to water scarcity and water demand conflicts by competing for water and land with food production (FAO, 2008). Increased water scarcity in the future might push countries to invest in desalination facilities. However, desalination requires more energy than traditional water treatment. Therefore, a future with more desalination plants would increase energy needs and GHG emissions by the water sector, unless renewable energy is used (World Bank, 2012b).

Time, place and quality of abstractions and releases: Traditional once-through cooling systems withdraw large amounts of water (with detrimental effects on water ecosystems) but return most of the water to the source, making it available for additional uses downstream. Cooling towers withdraw considerably less water but consume almost all of it, reducing the water available downstream. The shift from once-through to cooling towers would reduce water withdrawals per unit of energy produced, but increase the water consumed (Wolfe *et al.*, 2009; IEA, 2012a; Delgado Martin, 2012). The water discharged by a user or operation determines its future use. In some cases, the quality of the water returned to the environment may be a hindrance or prohibitive to another sector's use. Additionally, the timing of water resources is critical, as energy operations are greatly determined by water availability in time and place.

National perspective vs. regional perspective: Water issues are very region-specific, and often global or national studies, data aggregations and analyses are not sufficient to understand the magnitude of the issues locally. For example, the total percent of water used by the oil and gas industry at the national level might reflect that the percentage is very small, but at a regional level that percentage may be significant, revealing that continued water management practices may have adverse impacts on water resources, the environment and the long-term viability of energy operations in that area.

Life cycle trade-offs: For example, combined cycle power plants are more efficient, require less water and emit less GHGs than many other power plants. However, if natural gas extracted through hydraulic fracturing (fracking) fuels the power plant, the overall picture could change. Fracking requires large amounts of water and generates wastewater that needs to be treated to avoid adverse impacts on the environment. Moreover, if operations are not done properly fracking can pollute local water reserves and release significant amounts of methane, a potent GHG. Thus, depending on local conditions and management of energy extraction and the regulatory framework, the water gains obtained in electricity generation might come at the expense of increased water use and diminish the quality of water resources during the extraction process.

Water for energy vs. water for food: One of the more difficult issues to manage is the fact that the economic value of water to the energy sector, at the margin, will generally be greater than its economic value to agriculture, while the implicit political power of the agricultural sector can sometimes be greater than that of the energy sector. This implies that the energy sector will generally be willing and able to pay more for water than competing agricultural uses. The risk associated with this is that some agricultural groups may seek to use their political power to redress this difference in economic power, such as by portraying the energy sector as damaging agricultural interests and threatening food security.

All these complexities and unintended consequences are important in the policy-making process. These and other trade-offs, such as the potential conflicts with other water users, particularly in water-scarce regions and basins, need to be properly analyzed to be able to understand the long-term sustainable impacts of different water and energy plans and investments.

Solutions

There are many opportunities to increase energy access and the share of renewables and sustainably increase water-use efficiency in the energy sector, in order to both adapt to and mitigate climate change (Arent *et al.*, 2014). Chiefly, demand and supply of water and energy can be better managed through more efficient use of both resources. It is important to take into account that some adaptations can come at a mitigation cost, such as building more robust infrastructure that is climate-resilient but results in larger emissions of greenhouse gases (WBCSD, 2009).

One way to reduce freshwater use in the energy sector is by fostering the use of water-efficient cooling systems such as dry cooling in thermal power plants (Maulbetsch, 2011), using non-freshwater resources such as seawater or wastewater for cooling (Argonne National Laboratory, 2007; US DOE, 2009), recycling and reusing water in energy extraction facilities, focusing on the development of renewables such as wind or solar PV that require almost no water to produce electricity (Arent *et al.*, 2014), or developing sustainable multipurpose dams. However, there are trade-offs. For example, dry-cooled systems can cost 2 to 16 percent more than closed loop cooled systems (Maulbetsch and DiFilippo, 2006; CEC, 2002) and also decrease the efficiency of the power plant, thereby increasing GHG emissions. Therefore, dry-cooled systems are best suited to wetter and cooler climates where they can operate at nearly the same efficiency as wet-cooled systems. However, it is usually in dry climates where their use is needed the most.

However, there are also solutions, such as increasing energy and water efficiency (Cooper and Sehlke, 2012; Wallis *et al.*, 2014). One approach to enhance energy efficiency in the supply side is to shift from old, inefficient power plants to new, more efficient models, producing the same energy output with less fuel and water input and thereby saving energy and water and decreasing GHG emissions. Similar benefits would be seen through energy efficiency gains on the demand side. Another means to save water and energy can be achieved in the water sector, for example, by preventing leaks and repairing damaged systems. In Vietnam the World Bank financed a performance-based leakage contract in Ho Chi Minh City. Leakage was reduced from 54 percent to 29 percent. This helped save around 92,000 cubic metres of water and 23,000 kilowatt hours of electricity per day through reduced pumping. That is the equivalent of thirty-six Olympic-sized pools of water per day, and enough power for a community of 27,000 people in rural Vietnam (with average electricity consumption of 2009). Furthermore, water and wastewater treatment plants could dramatically save energy and reduce operational expense by increasing efficiency. In water and wastewater utilities, electricity costs are usually between 5 to 30 percent of total operating costs (in developing countries it can reach up to 40 percent). Given such energy costs, measures to improve energy efficiency often have a payback period of less than five years (ESMAP, 2012). Thus, investing in energy efficiency can improve the financial health of the plants and enable quicker and greater expansion of clean water access for the poor by making the system cheaper to operate (ESMAP, 2013). Wastewater treatment plants may also consider selling their treated wastewater as an alternate water source to nearby power plants or other possible users to ensure an additional and stable revenue stream. In San Luis Potosi, Mexico, new water reuse regulations and a creative project funding contract (Built Own Operate Transfer – BOOT) incentivized wastewater reuse (Lazarova *et al.*, 2013). Instead of using freshwater, a power plant uses treated effluent from a nearby wastewater treatment plant in its cooling towers. This wastewater is 33 percent cheaper than groundwater, which has resulted in US$18 million of savings for the power utility (over six years). This extra revenue covers almost all the operation and maintenance costs of the wastewater treatment plant. Moreover, groundwater extractions have been reduced by 48 million cubic metres in six years. This project highlights how innovative, integrated approaches can be of both environmental and economical benefit.

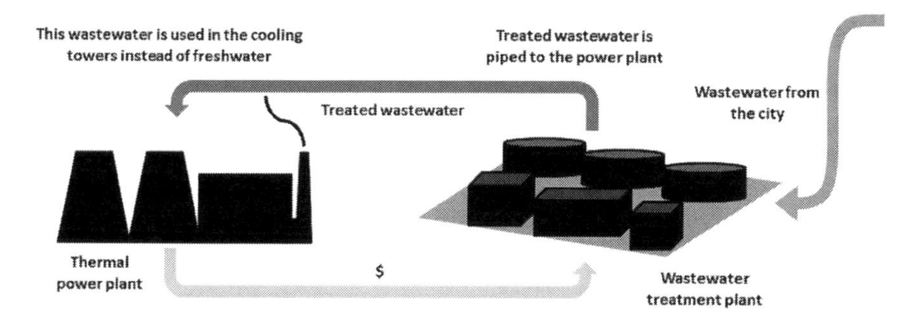

Figure 13.4 Diagram of Project Tenorio Innovative solution to reduce groundwater over extraction (based on Thirsty Energy, World Bank: www.worldbank.org/thirstyenergy. For more information see also http://www.reclaimedwater.net/data/files/240.pdf).

In addition to the pursuit of new technical solutions, new political and economic frameworks need to be designed to promote cooperation and integrated planning among sectors. All solutions depend on many factors, including technology, environment and financial, and they are site-specific. Integrated investment decisions and planning can ensure that all the complexities of energy and water are properly addressed (Rodriguez *et al.*, 2013). For example, incorporating water costs and physical constraints and availability in energy planning models can ensure more sustainable energy development. Economic frameworks can be used to quantify impacts on energy sector investments, the broader economy, and understanding economic, social and environmental trade-offs among competing water uses. These approaches better capture the complexity of water and energy issues and establish a systematic approach to take into account all the existing interactions and relationships between sectors. Comprehensive approaches that consider the diverse set of factors that influence energy and water demand and incorporate those issues into solutions will provide a robust management framework for the Water–Energy Nexus. Management capacities will be strengthened by integrated modelling approaches that allow governments to adapt to change, such as population and economic growth. This will enhance a nation's resiliency in the face of uncertainties brought on by climate change.

Conclusions

Meeting future demands for water and energy resources under climate change uncertainty requires innovative approaches that encourage cross-sectoral cooperation and improve analysis of water and energy trade-offs at the national and regional levels. Climate change can aggravate existing pressures and increase future uncertainties, and these uncertainties must be clearly incorporated in planning and investment design to ensure that infrastructure investments are sustainable and resilient.

Water is crucial for energy production, and water constraints could limit our energy options in the future. Thus, as climate change mitigation policies and strategies are designed, it is important to not only evaluate energy costs and greenhouse gas (GHG) mitigation potential, but also to assess the impact of those policies on water resources. Some mitigation options might not be as attractive if water impacts and costs are considered. For example, increasing the share of biofuels in renewable energy generation for climate change mitigation may have a greater impact on water resources than other renewables. We will only be able to achieve sustainable energy for all if we evaluate and consider technologies and policies that can reduce GHG emissions, water resource and environmental impacts, and energy costs.

Solutions need to be contextualized and localized for enduring success. For example, forcing the energy sector to reduce its water withdrawals everywhere in the world without first analyzing the water stress of the region and future water availability could increase the cost of energy production in areas where water for energy is not a concern. On the other hand, infrastructure choices and decisions made today about which extraction facilities to develop, which power plants to build, which to retire, and which energy or cooling technologies to deploy and develop matter. Energy and water infrastructure is designed to last for decades, and thus decisions should take into account future water availability, including a consideration of climate change impacts and increasing future competing water demands such as for agriculture, municipal and industrial uses. Energy and water infrastructure and operators should equally consider the likelihood of extreme weather events and their exposure to those threats. As energy planners consider these risks, they will help to ensure that the dependence of the energy sector on water does not compromise water resources, and that expensive retrofitting is avoided while a more water-smart energy sector is developed.

Coordination and collaboration across institutions must be strengthened. Even though there is acknowledgement of the interconnectedness of the two sectors and the potential impact of climate change in a world of rapid technological advancement and economic and population growth, there are inadequacies in existing policies and institutions. In most developing countries, the existing institutional fragmentation is an impediment to ensuring that planning and investments are done in a coordinated manner. As such, we have public policies, legal and regulatory frameworks that do not identify and address the trade-offs across energy and water.

Defining goals with energy and water interactions in mind will work towards a real sustainable development; we must cut across sectors in our thinking as improvements for one goal may have negative impacts on another. As we consider goals for the future, we can act today by employing best practices that enhance efficiency and integrate planning.

These approaches will help governments and companies avoid financial losses in energy and power investments, infrastructure prone to risk in the wake of climate change, and unstable economies. As regions of the world continue to experience water constraints and pressures mount due to climate change and increasing populations, innovative solutions will be crucial to ensure the sustainable development of water and energy resources.

Notes

1 Open cycle power plants (mainly used as peak power plants) do not use the steam cycle to turn the turbines, and hence do not require water for cooling.

2 From a regional water consumption perspective, non-cooling plant processes are usually negligible. However, these streams of water contain several pollutants and should be treated before being returned to the water source, otherwise they can have a negative impact on the environment. Thus, from a plant-level economic standpoint, there may be very significant costs related to treating wastewater for discharge under stringent regulations.

References

Arent, D., Pless, J., Mai, T., Wiser, R., Hand, M., Baldwin, S., Heath, G., Macknick, J., Bazilian, M., Schlosser, A. and Denholm, P. (2014) 'Implications of high renewable electricity penetration in the US for water use, greenhouse gas emissions, land-use, and materials supply', *Applied Energy*, 123(15), 368–377.

Argonne National Laboratory (2007) *Use of Reclaimed Water for Power Plant Cooling*. For the US Department of Energy by J. A. Veil, Environmental Science Division, Argonne National Laboratory, ANL/EVS/R-07/3.

California Energy Commission (CEC) (2002) *Appendix A to Water and Soil Resources. SMUD Cosumnes Power Plant. Water Supply and Cooling Options*, Sacramento: CEC.

CDP (Carbon Disclosure Project) (2013) *CDP Global Water Report 2013. A need for a step change in water risk management*. Available online at: https://www.cdp.net/CDPResults/CDP-Global-Water-Report-2013.pdf

Chandel, M.K., Pratson, L.F. and Jackson, R.B. (2011) 'The potential impacts of climate-change policy on freshwater use in thermoelectric power generation', *Energy Policy*, 39(10), 6234–6242.

Cooper, C.C. and Sehlke, G. (2012) 'Sustainability and energy development: Influences of greenhouse gas emission reduction options on water use in energy production', *Environmental Science and Technology*, 3509–3518, Washington, DC: American Chemical Society. Available online at: http://pubs.acs.org/doi/abs/10.1021/es201901p

Delgado Martin, A. (2012) *Water Footprint of Electric Power Generation: Modeling its use and analyzing options for a water-scarce future*, Master's thesis, Massachusetts Institute of Technology.

DiPietro, P., Gerdes, K. and Nichols, C. (2009) *Water Requirements for Existing and Emerging Thermoelectric Plant Technologies*, Technical Report No. DOE/NETL-402/080108, Pittsburgh, PA: US DOE National Energy Technology Laboratory.

Electric Power Research Institute (EPRI) (2010) *Cooling Requirements and Water Use Impacts of Advanced Coal-fired Power Plants with CO_2 Capture and Storage*. Interim Results, Palo Alto: EPRI. Available online at: www.epri.com/abstracts/Pages/ProductAbstract.aspx?ProductId=000000000001024495

ESMAP (2012) *A Primer on Energy Efficiency for Municipal Water and Wastewater Utilities*, Technical Report, Energy Sector Management Assistance Program, Washington, DC: World Bank.

ESMAP (2013) *A Primer on Energy Effi ciency for Municipal Water and Wastewater Utilities*, Technical Report, Energy Sector Management Assistance Program, Washington, DC: World Bank.

Food and Agriculture Organization (FAO) (2008) *Climate Change, Water and Food Security*. Technical Background Document from the Expert Consultation, February 26–28, Rome, Italy: FAO.

Food and Agriculture Organization (FAO)-OECD (2012) Chapter 3: Biofuels, in FAO-OECD *Agricultural Outlook 2012*, Rome: FAO-OECD. Available online at:. http://www.fao.org/fileadmin/templates/est/COMM_MARKETS_MONITORING/Oilcrops/Documents/OECD_Reports/biofuels_chapter.pdf

Goldstein, R. and Smith, W. (2002) *Water and Sustainability (Volume 4): US Electricity Consumption for Water Supply & Treatment – The Next Half Century*. Tech. Rep., Palo Alto, CA: Electric Power Research Institute.

IEA (International Energy Agency) (2012a) 'CO_2 emissions by product and flow', IEA CO_2 Emissions from Fuel Combustion Statistics (database), Paris: IEA. Available online at: www.iea.org/statistics/

IEA (International Energy Agency) (2012b) *World Energy Outlook 2012*. Paris: International Energy Agency.

IEA World Energy Outlook, 2013.

IPCC (2013) 'Summary for Policymakers', in T.F. Stocker, D. Qin, G.-K. Plattner, M. Tignor, S.K. Allen, J. Boschung, A. Nauels, Y. Xia, V. Bex and P.M. Midgley (eds), *Climate Change 2013: The Physical Science Basis. Contribution of Working Group I to the Fifth Assessment Report of the Intergovernmental Panel on Climate Change*, Cambridge, UK and New York, USA: Cambridge University Press.

Kenny, J.F., Barber, N.L., Hutson, S.S., Linsey, K.S., Lovelace, J.K. and Maupin, M.A. (2009) 'Estimated use of water in the United States in 2005', *US Geological Survey Circular*, 1344, 52.

Lazarova, V., Asano, T., Bahri, A. and Anderson, J. (2013) *Milestones in Water Reuse: The Best Success*, London: IWA Publishing.

Macknick, J., Newmark, R., Heath, G. and Hallet, K.C. (2011) *A Review of Operational Water Consumption and Withdrawal Factors for Electricity Generating Technologies*, Technical Report No. NREL/TP-6A20-50900. Boulder, CO: US DOE National Renewable Energy Laboratory.

Maulbetsch, J.S. (2011) *Cost/Performance Comparisons of Water Conserving Power Plant Cooling Systems*. Paper No. IMECE2011-63135, American Society of Mechanical Engineers (ASME) Conference Proceedings, 385–390.

Maulbetsch, J.S. and DiFilippo, M.N. (2006) *Cost and Value of Water Use at Combined-Cycle Power Plants*, Sacramento: California Energy Commission, PIER Energy-Related Environmental Research, CEC-500-2006-034.

Newmark, R.L. *et al.*, (2010) 'Water challenges for geologic carbon capture and sequestration', *Environ. Manage.*, 45(4), 651–661.

Pate, R., Hightower, M., Cameron, C. and Einfeld, W. (2007) *Overview of Energy-Water Interferences and the Emerging Energy Demands on Water Resources*. Sandia National Laboratories, Albuquerque, New Mexico.

Rodriguez, D.J., Delgado, A., DeLaquil, P. and Sohns, A. (2013) *Thirsty energy*. Water papers. Washington DC: World Bank. Available online at: http://documents.worldbank.org/curated/en/2013/01/17932041/thirsty-energy

Rubbelke, D. and Vogele, S. (2011) 'Impacts of climate change on European critical infrastructures: The case of the power sector', *Environ. Sci. Policy*, 14, 53–63.

Stewart, R., Wollheim, W., Miara, A., Vorosmarty, C., Fekete, B., Lammers, R. and Rosenzweig, B. (2013) *Horizontal cooling towers: riverine ecosystem services and the fate of thermoelectric heat in the contemporary Northeast US*. Environmental Research Letters, Bristol: IOP Publishing. Available online at: http://iopscience.iop.org/article/10.1088/1748-9326/8/2/025010/pdf

Stillwell, A.S., King, C.W., Webber, M.E., Duncan, I.J. and Hardberger, A. (2011) 'The Energy-Water Nexus in Texas', *Ecology and Society*, 16(1), 2.

Stockholm Environment Institute (SEI) (2011) *Understanding the Nexus*, background paper for the Bonn 2011 Nexus Conference: The Water, Energy and Food Security Nexus, Solutions for the Green Economy, 16–18 November 2011.

Times of India (2014) *Parli thermal power station shuts down*. Available online at http://timesofindia.indiatimes.com/city/aurangabad/Parli-thermal-power-station-shuts-down/articleshow/39456130.cms

Thompson, A. and Climate Central (2014) '2014 on Track to Be Hottest Year on Record', *Scientific American*, September 23. Available online at: http://www.scientificamerican.com/article/2014-on-track-to-be-hottest-year-on-record/

UN (United Nations) (2010) *Population Facts*. Department of Economic and Social Affairs. Population Division. Available online at: http://www.un.org/en/development/desa/population/publications/pdf/popfacts/popfacts_2010-5.pdf

UNEP (United Nations Environmental Programme) (2010) *Water-related materiality briefings for financial institutions*, Issue 2: Power Sector, Geneva: UNEP Finance Initiative.

UNEP (United Nations Environmental Programme) (2012a) *Water-related materiality briefings for financial institutions*, Issue 3: Extractives Sector, Geneva: UNEP Finance Initiative.

UNEP (United Nations Environmental Programme) (2012b) *The UN-Water Status Report on the Application of Integrated Approaches to Water Resources Management*, Geneva: UNEP Finance Initiative.

US DOE (United States Department of Energy) (2009) *Use of Non-Traditional Water for Power Plant Applications: An Overview of DOE/NETL R&D Efforts*. DOE/NETL-311/040609. Washington, DC: DOE/NETL. Available online at: http://www.netl.doe.gov/File%20Library/Research/Coal/ewr/water/Use-of-Nontraditional-Water-for-Power-Plant-Applications.pdf

US DOE (United States Department of Energy) (2010) *Water Vulnerabilities for Existing Coal-fired Power Plants*. DOE/NETL-2010/1429. Washington, DC: DOE. Available online at: www.ipd.anl.gov/anlpubs/2010/08/67687.pdf

US DOE (United States Department of Energy) (2013) *US Energy Sector Vulnerabilities to Climate Change and Extreme Weather*. Washington, DC: DOE. Available online at: http://energy.gov/sites/prod/files/2013/07/f2/20130710-Energy-Sector-Vulnerabilities-Report.pdf

USGS (United States Geological Survey) (2013) *Water Resources and Shale Gas/Oil Production in the Appalachian Basin – Critical Issues and Evolving Developments*, Ithaca, NY: USGS. Available online at: http://pubs.usgs.gov/of/2013/1137/

van Vliet, T.H.M., Yearsley, R.J., Ludwig, F., Vögele, S., Lettenmaier, P.D. and Kabat, P. (2012) 'Vulnerability of US and European electricity supply to climate change', *Nature Climate Change*, 2, 678–681. Available online at: http://www.nature.com/nclimate/journal/v2/n9/full/nclimate1546.html

van Vliet *et al.*, (2013) 'Water constraints on European power supply under climate change: impacts on electricity prices', *Environ. Res. Lett.*, 8, 035010.

Wallis, P.J., Ward, B.M., Pittock, J., Barnsey, H., Denis, A., Kenway, J.S., King, W.C., Mushtaq, S., Retamal, L.M. and Spies, R.S. (2014) 'The water impacts of climate change mitigation measures', *Climatic Change*, 125(2), 209–220. DOI: 10.1007/s10584-014-1156-6. Available online at: http://link.springer.com/article/10.1007%2Fs10584-014-1156-6

Webber, M. (2008) 'Water versus Energy', *Scientific American*, Earth 3.0, 34–41.

WHO/UNICEF (2014) *Progress on Drinking Water and Sanitation – 2014 update*, Geneva, Switzerland: World Health Organization.

Wolfe, J.R., Goldstein, R.A., Maulbetsch, J.S. and McGowin, C.R. (2009) 'An electric power industry perspective on water use efficiency', *Journal of Contemporary Water Research and Education*, 143, 30–34.

World Bank (2012a) *Turn down the heat: Why a 4°C warmer world must be avoided*. Washington DC: World Bank. Available online at: http://documents.worldbank.org/curated/en/2012/11/17097815/turn-down-heat-4 percentC2 percentB0c-warmer-world-must-avoided

World Bank (2012b) *Renewable Energy Desalination: An emerging solution to close the water gap in the Middle East and North Africa.* MENA Development Report. Washington DC: World Bank.

World Business Council for Sustainable Development (WBCSD) (2009) *Water, Energy and Climate Change: A contribution from the business community*, Switzerland: WBCSD.

World Economic Forum (2011) *The Water-Energy Nexus: Strategic Considerations for Energy Policy Makers*, Zurich: WEF. Available online at: www3.weforum.org/docs/GAC/2014/WEF_GAC_EnergySecurity_WaterEnergyNexus_Paper_2014.pdf

World Energy Council (WEC) (2010) *Water for Energy*, London, UK: World Energy Council.

World Resources Institute (WRI) (2011) *Aqueduct and the Water-Food-Energy Nexus.* Available online at: http://insights.wri.org/aqueduct/2011/11/aqueduct-and-water-food-energy-nexus

WWAP (United Nations World Water Assessment Programme) (2012) *The United Nations World Water Development Report 4: Managing Water under Uncertainty and Risk.* Paris: UNESCO.

WWAP (United Nations World Water Assessment Programme) (2014) *The United Nations World Water Development Report 2014: Water and Energy*, Paris: UNESCO.

Part 5

Nexus perspectives

Food, water, and climate

14 Smallholder farmers are at the Nexus of post-2015 development issues

Iain MacGillivray

Eradicating the scourge of poverty, hunger and malnutrition in all its forms is at the heart of a new and universal post-2015 agenda. In "leaving no one behind", rural–urban inequalities must be addressed – with particular attention to small-scale food producers including women, indigenous peoples and family farmers. Really, this is a defining moment for the International Fund for Agricultural Development (IFAD) as these issues reaffirm its mandate and take centre stage in the post-2015 debate.

Investing in rural people is IFAD's business. It recognizes that the women, men and children in developing countries who depend on smallholder agriculture, forestry, livestock and fisheries are the custodians of vital natural resources and biodiversity, and are central to climate change mitigation and adaptation. It also contends that smallholder agriculture development and rural transformation need to be an integral part of the post-2015 global development agenda, if that agenda is to succeed. Three-quarters of the world's poor and hungry people live in the rural areas of developing countries. The crafting of a new development agenda, and its implementation, is a unique opportunity to refocus policy, investments and partnerships on inclusive and sustainable rural transformation.

Without such transformation, rural–urban inequalities will only deepen. Cities will struggle and global food security will be at risk. Conversely, rural transformation and the growth of the rural economy can be a powerful engine of inclusive sustainable development in all its aspects, from economic growth and employment to poverty eradication, from a healthy environment to inclusive societies, from gender equality to food and nutrition security for all.

Although they do not cover the whole rural development agenda, four clusters of issues provide a map of areas where catalytic action could be inspired by the new goals, targets and indicators, adapted to individual country contexts.

First among the four key issues that IFAD is emphasizing in its support for the process of evolving a new agenda – reflecting IFAD's position on the post-2015 agenda as indicated in its four policy briefs available online – is the possibility of leveraging the rural–urban Nexus for development. The second issue focuses on developing an empowerment agenda for rural people, which would concentrate on the social and political inclusion challenges that impede progress towards rural transformation, rural–urban integration, smallholder agriculture, growth and resilience. The third issue is the imperative to focus on agriculture, food and nutrition security – specifically by investing in smallholder agriculture, the main source of food in many developing countries. The last issue is the necessity of promoting the resilience and sustainability of rural livelihoods.

Leveraging the rural–urban nexus of development

The mistaken view of rural areas as a synonym for backwardness, marginality and poverty obscures the potential of rural people and their communities. The perception that the only function of the rural space is to shed labour in order for this labour to feed the urban slums and growing cities – assuming that these urban slums can really work in isolation from the rest – is mistaken and needs to change. We should not overlook who will feed the urban populations and provide needed environmental services, which in turn depend on how natural resources are managed in rural areas. We should not delude ourselves: if migration is going in one direction, remittances are going in the other, and there is a whole 'missing middle' composed of small towns and industries, where supply chains are formed and landscapes managed; these represent a continuum between rural and urban areas. Here, real osmosis is occurring – a constant, often intuitive yet boundless exchange of knowledge and ideas through continual exposure in the rural–urban Nexus. This mutual interdependence is very real, though it remains to be better acknowledged and nurtured – and any implementation of the post-2015 framework must recognize this.

Can we afford a world where cities are assumed to be sustainable but the countryside is not? Sustainable cities require sustainable rural areas. By the same token, can we risk or afford a future where rural–urban inequalities keep increasing, polarized societies are undermined, families are divided (where both men and women migrate, often leaving their elders to mind their young), with disgruntled unemployed youth, social unrest, poor schooling, degraded environments, pockets of poverty, no rural infrastructure and insufficient food supply? This cannot possibly be the future we want.

The future we want is a future that genuinely supports a sustainable rural–urban Nexus, where the focus is more on rural towns and local governments, where it is possible to work on the farm and live in nearby towns. This will require investments, infrastructure, services and more financial inclusion (especially through information and communication technologies or ICTs), and leveraging the huge amounts of remittances that are going back to rural areas from urban centres. It is also the 'space' where food security and supply chains will be generated which will feed the cities and potentially absorb a greater number of the youth and unemployed, providing them with improved prospects to stay in rural areas with gainful employment and opportunities. We have seen the limitations of cities and their capacity to employ youth. Clearly, rural areas have to be better used and developed to contribute and thereby tackle these inadequacies. This is an area where we can speak about sustainable landscapes and at the same time about sustainable 'peoplescapes', because the two really go together.

Promoting an empowerment agenda for rural livelihoods

The five hundred million smallholder family farms in the world support 2 to 2.5 billion people – and most of these are poor. About eight hundred million people are undernourished. We know that 75 percent of extreme poverty is rural, 25 percent is urban. We also know that agricultural growth is at least twice as efficient in reducing poverty compared with other types of growth, and we know that agriculture in the developing world is mostly a smallholder business. If you examine this equation, the solution is that improving smallholder family agriculture is the closest proxy available for reducing global poverty and hunger. Figure 14.1 illustrates the numbers of people living in extreme poverty in both rural and urban areas.

IFAD believes this signifies the need for an empowerment agenda – social and economic empowerment. Through social empowerment, people own their organizations to become actors for collective action, building social capital and trust at the community level by means such as securing rights and access to services and assets as well as respect and equality for all, including across gender and ethnicity lines. Under economic empowerment, the agenda involves ensuring people's access to productive assets – land, livestock, equipment and technology. Financial inclusion, whether through savings, credit, insurance or remittances, enables aggregation and going to scale. This is an important agenda.

Voice and scale will help to enable farm producers and rural people to express the best of their capacities to negotiate with government and the private sector. For this, 'institutions of rural people' will be critical (as opposed to institutions for rural people). These embrace three main entry points:

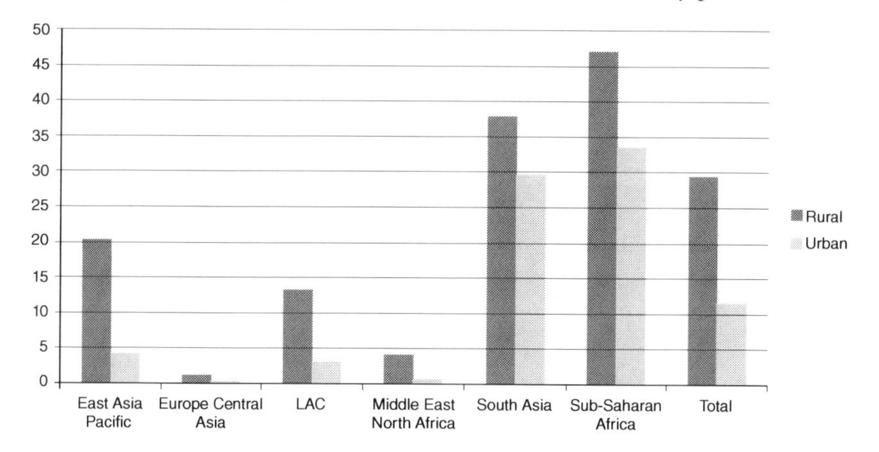

Figure 14.1 People in both rural and urban areas living in extreme poverty.

Source: Global Monitoring Report 2013 – Rural–Urban Dynamics and Millennium Development Goals.

1. affinity-based savings and loans group (for example, self-help groups, women's groups etc.)
2. asset and market type groups (in the form of cooperatives of commodity-based groupings)
3. resource-based groups (like water user associations and watershed committees).

Ultimately, the aim is to bring people together, supporting their organizations to make them real and effective – in essence, giving rural individuals and communities voice, notably those living in poverty, and supporting them to empower themselves at the appropriate scale.

Investing in smallholder agriculture for global food security and nutrition

The farms in the developing world that support the livelihoods of close to a third of the world's population are mostly managed by poor smallholders, many of whom are women. They suffer from poor quality diets and malnutrition due to inadequate consumption from their own production. They regularly lack access to other sources of food. In spite of their situation of an often permanent state of food and nutrition insecurity, smallholder farmers are responsible for about 80 percent of the food produced in sub-Saharan Africa and parts of Asia. Clearly, optimizing their contribution to agricultural production and to food systems at large, through better health and nutrition, is of great consequence. The question looming on the horizon is who will help to feed the more than nine billion people who are projected to inhabit the world by 2050 – an increase in excess of two billion people over today's population.

At IFAD, we believe this is not going to happen without the contribution of smallholder family farms. It will also be possible because of the revival of the agricultural sector that has benefitted from increasing prices since 2008, offering new opportunities within the sector. The private sector is clearly on the move and is expanding its investments and partnerships rapidly – in fact, many see this as the end of an era of cheap food. While posing challenges for food-insecure farmers and consumers, especially those who spend a large proportion of their disposable income on food, better prices offer an opportunity to contribute to reducing poverty and global food security through market-oriented smallholder farming.

To do this, a more comprehensive approach will be needed to optimize agriculture's contribution to good nutrition and make food systems nutrition sensitive – by ensuring that nutrition outcomes for rural and urban people are taken into account both at the design stage and during the implementation of agricultural investments. In the quest for increasing agricultural productivity in a sustainable way, agriculture must also strengthen food security and deliver environmental benefits that bolster farmers' resilience, reduce agriculture's greenhouse gas emissions and increase carbon sequestration, thereby making the sector climate sensitive.

How is it possible for the private sector to be willing and interested in engaging with the smallholder community? The answer lies in what is referred to as business at the bottom of the pyramid. If the pyramid in Figure 14.2 represents smallholder populations, we know that the private sector is engaging further down the pyramid. So far, they've been able to engage with the commercial segment of that pyramid. They clearly want to go deeper, but the more they do so the harder it gets, because transaction costs and associated risks increase.

We also know that the public sector is going up the pyramid, working mostly with subsistence agriculture (admittedly while also working with farmers who are in one way or other connected to markets). It is in the middle that we find the viability gap where we would like to attract the private sector, even if we know this is not going to be obvious or intuitively easy. IFAD is convinced this is the biggest untapped market, representing the largest potential for productivity gains. Undoubtedly it remains a kind of special market because it is a low-cost, low-profit, high-volume market. This is where aggregation and scale are important. Without scale, it is extremely hard for the private sector to go down this pyramid because they need volume. Without volume, it is almost impossible to do business with small, isolated and dispersed farmers.

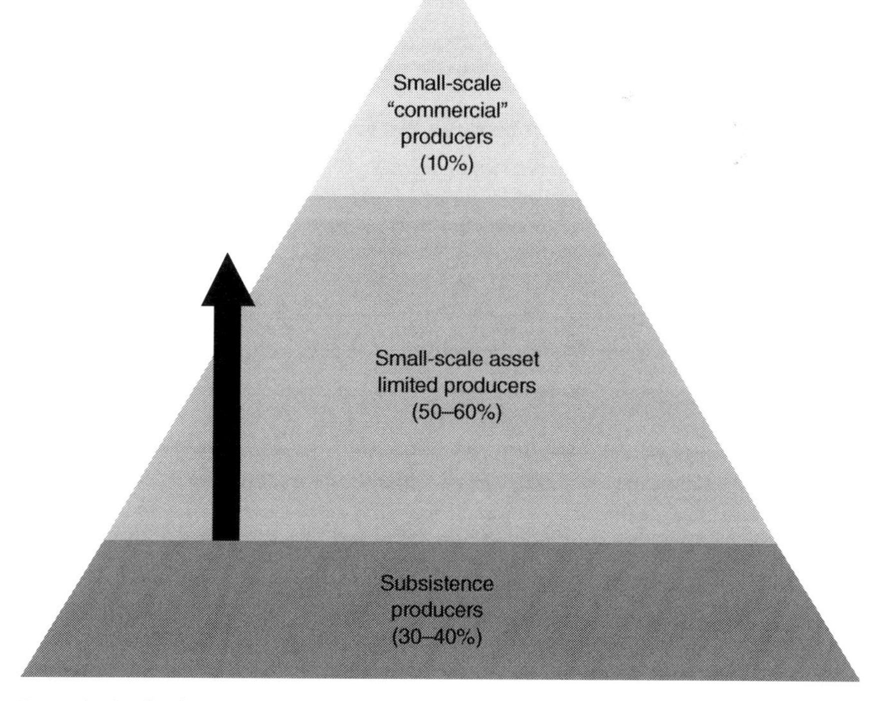

Figure 14.2 The five-hundred-million smallholder pyramid.
Source: International Fund for Agricultural Development (IFAD).

This is really the five-hundred-million smallholder pyramid. It represents a huge market of over two billion people who are not only producers but also consumers. So what can be done to fill this viability gap? IFAD works through governments to help reduce the transaction costs of doing business – largely by organizing and training farmers into producer groups. IFAD also provides finance to the public and semi-public sectors through rigorous business plans, in order to encourage the formation of public–private producer partnerships between the private sector and farmers. IFAD also leverages supply chain financing, where the private sector is encouraged to embrace a financial inclusion agenda by investing themselves in smallholder business operations.

Promoting the resilience of rural households

Smallholder farmers are key to promoting the resilience of rural households because they are the managers of a large part of the world's natural resources: land, water, forests and fish. They are affected by climate and in need of energy. Because of this, the search for Nexus solutions will need to involve smallholder farmers.

Unfortunately, smallholder farmers are both unwilling victims of and contributors to climate change. Agriculture and the environment have been both friends and foes, working at cross-purposes for a long time. To grow (food), agriculture needs environmental resources that we also want to protect. Hence there is a huge agenda of trade-offs, but also of unexploited synergies and opportunities, since it is still possible to have win–win solutions where yields can be increased and natural resources managed sustainably. We know of the climate risks, and we have started to understand the impact of food supplies on food security because we know that shocks to the food supply system can have multiplier effects on price shocks. Conflict and poor governance also affect smallholder farmers. For example, the poor are particularly vulnerable to shocks. They can be severely affected by corruption and other factors as well, and resource losses affect their livelihood and can set them back for decades.

So the need for resilience is very real. We can cope with shocks and external factors through various means. We need to build resilience into grassroots institutions and strengthen their social capital and their institutions. We also need to do better risk assessments when we plan our interventions or when designing our projects, to be more conscious of the likely impact of possible climate or price shocks. We must be a bit smarter in terms of the innovation and technology that may be generated locally, recognizing that much of this technology may already exist. Better price and ecosystem risk management tools should be developed. Two examples of IFAD programmes that have been developed to address some of the various challenges are the Adaptation for Smallholder Agricultural Programme (ASAP), which is mainstreaming climate change throughout the portfolio, and the Platform for Agricultural Risk Management (PARM), which looks at risk management issues in a holistic way in selected countries in sub-Saharan Africa.

Sustainable agriculture needs to be seen from a multiple benefits point of view, as agriculture is much more than the sum of the calories it produces – it is also about landscapes, water, energy, lifestyle, social capital, nutrition and care of the environment. For all these reasons, we really need to look at sustainable agriculture differently and award it much higher priority than we have in the past.

Consolidating action for a better future

To create the world we want – a world where no child suffers chronic hunger, a world where all women and men have dignified jobs, a world where families sleep beneath a solid roof, a world where there is opportunity for all – we must develop our rural areas. IFAD sees rural transformation as a holistic approach that does just this, looking at all the factors involved in revitalizing the rural space in a sustainable, equitable and inclusive manner.

Sustainable smallholder agriculture provides a powerful vehicle to deliver multiple benefits across all of the dimensions of sustainable development for people no matter where they live. While needing to deliver global food security and improved nutritional outcomes, sustainable smallholder farming also offers expanded employment opportunities while managing the natural resource base sustainably and remaining resilient to the effects of climate change. It is not peripheral, but rather at the very nexus of sustainable, inclusive development.

While this agenda brings significant challenges, it also offers new opportunities to smallholders through expanding markets for food and environmental services, as well as new sources of growth and employment in larger rural economies. As the international community's most direct conduit for channeling investment to smallholder family farmers, IFAD is giving explicit attention to innovation, learning and scaling up. By intensifying its efforts to assist smallholders to adapt to climate change, improve nutrition outcomes and further its reach on gender equality and women's empowerment, IFAD aims to achieve transformative impacts on rural areas. Consolidating its approaches to public–private-producer partnerships, country-level policy engagement, global policy engagement, and South–South and triangular cooperation are key elements of its agenda.

These are the kinds of steps that IFAD will be taking and the outlook that it is promoting among its partners and the development community as a whole. For IFAD, a future where smallholder family farming is at the centre of the agricultural, economic, environmental, and social agendas is essential for promoting equitable and sustainable development.

15 Green opportunities for urban sanitation challenges through energy, water and nutrient recovery

Pay Drechsel and Munir A. Hanjra

Introduction

A Nexus approach means that when governments and industries determine policies in one sector – whether energy, agriculture or water – they take into account the implications in other sectors (Bird, 2014). Applying this approach within an urbanizing world gives the Nexus a vital justification, since nowhere else are the flows of energy, water and food so stressed as where populations peak within confined spaces. Van Rooijen *et al.* (2005) coined the term 'Sponge City' to visualize the urban metabolism which is 'sucking' resources from its periphery, while usually releasing a less appreciated return flow. In fact, cities are hungry and thirsty and enormous hubs of consumption of all kind of goods, including food, water and energy. This in turn makes them major centres of solid and liquid waste generation. If this waste remains in the urban area or its landfills, cities will also become vast sinks for valuable resources, like crop nutrients, while rural production areas increasingly face soil fertility degradation. Urban waste is today not only the paramount environmental and health challenge that growing cities face, but also a significant economic challenge in those countries where waste collection and treatment cannot be financed through taxes and fees, thus questioning the sustainability of urban growth (Villarroel Walker *et al.*, 2012).

In this chapter we summarize findings from West Africa which analyzed the metabolism of four cities, visualizing the challenges of linear resource flows for system sustainability, followed by a discussion of options for transforming the urbanization challenge into an opportunity for a circular economy.

Supporting a move towards a circular metabolism where resources are used again and again within the context of a circular economy offers opportunities for increasing urban resilience and value creation from the recovery and reuse of resources that would otherwise be irretrievably lost. This also presents urgently needed cost recovery options for waste management and sanitation, and business opportunities for private sector engagement (Otoo *et al.*, 2012). Resource recovery and reuse can ultimately lower the urban footprint on what are probably the most affected peri-urban ecosystems, and contribute significantly to sustainable urban growth within the rural–urban continuum.

Linking sanitation and business thinking

Sustainability demands that, for instance, wastewater is used as a renewable resource from which water, nutrients and energy can be recovered to support local needs including ecosystem services (Grant *et al.*, 2012). The traditional sanitation paradigm of treatment for safe disposal, which included the removal of potentially harmful nutrients from the wastewater, has too often been transferred to countries and situations where other sectors, in particular agriculture, would welcome reuse including the removed nutrients (Murray and Buckley, 2010; Guest *et al.*, 2009). In Guanajuato, Mexico, for example, the estimated cost for farmers of replacing nitrogen and phosphorus loss through improved wastewater treatment was estimated at US$900/ha (Scott *et al.*, 2000). There is thus an increasing call for a transition into post-modern sanitation systems linking agro-sanitation with business modelling (Ushijimaa *et al.*, 2014) to focus on what can be *recovered* and *reused*, rather than removed, to transform wastewater treatment and landfill facilities into resource recovery centres of the future. To realize a 'design for reuse' concept (Murray and Buckley, 2010) a stronger focus on the demand side for recovered resources is required; especially in low-income countries the common business model is still too supply driven (Box 15.1). A key incentive for leveraging private capital and private–public partnerships with players experienced in the agricultural input sector would be the extension of fertilizer subsidies for recovered nutrients. So far we see more of the opposite, where for health reasons the registration of recovered nutrients, e.g. struvite, is more cumbersome and costly than, for example, bringing an often less pure rock-phosphate on the market (Otoo *et al.*, 2015).

Box 15.1 Supply-side savings versus revenues from demand

In many African municipalities we see a strong interest in municipal waste composting, as it can significantly reduce the transport volume to a landfill and thus also the landfill lifetime. This matters, in particular due to increasing difficulties of finding space in urban proximity for landfills (communities fight the anticipated loss of land value) and the resulting transport costs to drive the waste far away from any populated areas. These costs are often up to two-thirds of the overall waste management budget. Compost stations, on the other hand, are easier to accept. If the compost is eventually burnt or becomes a second revenue stream it appears to be of much lower relevance (Drechsel *et al.*, 2010), especially since with increasing city size the gains from volume reduction might outpace those from compost market expansion.

This business model is also adopted in private–public partnerships when the private partner is based in the sanitation sector and runs waste collection and composting, but stops operations as soon as governmental payments for the service cease, as recently seen in Accra, Ghana (GNA, 2014). Compost sales are not a decisive factor here; in fact, in many cases the reuse market has not even been explored.

A similar situation can be found in wastewater treatment where phosphorus (P) recovery as struvite receives much attention. Modern technologies allow the recovery of high percentages of P before it starts damaging pipes and valves through unwanted precipitation. This results in significant savings for treatment operators by reducing the use of chemicals that would otherwise be needed to remove the crystals. Enterprises specializing in P recovery thrive on these savings, while the generated P fertilizer is still struggling to move beyond selected niche markets (Otoo *et al.*, 2015).

Sanitation service delivery and wastewater management decisions have traditionally been driven by consideration for public health, although full costs and benefits are infrequently included in the analysis and the public is rarely involved in decision making. In pursuit of sustainability, the goal should go beyond society and consider ecosystem service support, which is offered by resource recovery and reuse. This is obvious in view of the dwindling global P reserves (Rosemarin *et al.*, 2009). However, the lessons so far have also shown that closed loop processes do not manifest themselves through the promotion of ecological sanitation or the assumption that, for example, waste composting must be a win–win answer for farmers and waste managers as postulated optimistically more than a decade ago (Drechsel and Kunze, 2001). What is often described as an engineering challenge (*'Reinvent the Toilet'*) is increasingly understood as an institutional, social and economic challenge in dire need of profound business models, especially if scalability is targeted (Otoo and Drechsel, 2015). This requires market assessment, perception studies, health impact assessments and knowledge of technical options across scales which will feed into business modelling. In addition, an economic assessment of the planned interventions compared with the counterfactual 'business as usual' model is required to internalize any possible externalities, especially for human and environmental health, which could justify subsidies (Hanjra *et al.*, 2015; Mo and Zhang, 2013).

The urban metabolism

Urbanization is increasingly affecting inter-sectoral water allocations and trade, i.e. resource flows over large distances. A study of four cities in West Africa (Ouagadougou, Accra, Kumasi and Tamale) showed that Ghana's capital Accra receives its piped water from two basins, 80 percent of its food and the food-related virtual water from four sub-regional basins, and 20 percent of its food and virtual water from 38 basins worldwide, showing Accra's extensive water footprint to satisfy urban demands (Drechsel *et al.*, 2007). More interesting is the comparison of food flows and the urban supply from rural, peri-urban, and urban agriculture. With about a million tons of food entering the cities of Accra and Kumasi each year, of which a third (Accra) to a half (Kumasi) might leave the city boundaries again, the amount of food consumed and food waste generated in the cities is significant. Converting these food flows into nutrients, the quantity of nitrogen (N) that flows annually to Kumasi, for example, is more than the total amount of all annually imported N fertilizer in the whole of Ghana over several years (Drechsel *et al.*, 2007). As recycling activities are negligible, soils in agricultural production areas are mined of their fertility while the bulk of the food waste ends up either in landfills, street drains or the environment, making it evident that urban centres are indeed nutrient 'sinks' (Figure 15.1) with significant implications for environmental pollution and ecosystem resilience (Craswell *et al.*, 2004). That waste management cannot keep pace with urbanization was illustrated by data from Mensah (2004) and Vodounhessi (2006) showing that in the case of Kumasi, with its 1 million inhabitants at that time, about 18 percent of the generated solid waste and 66 percent of the faecal sludge remained uncollected in streets, pits and septic tanks.

While urban consumption is increasing with population growth, along with investments in water supply, investments in sanitation generally lag behind, and even more so the recovery of energy, water and nutrients from the urban

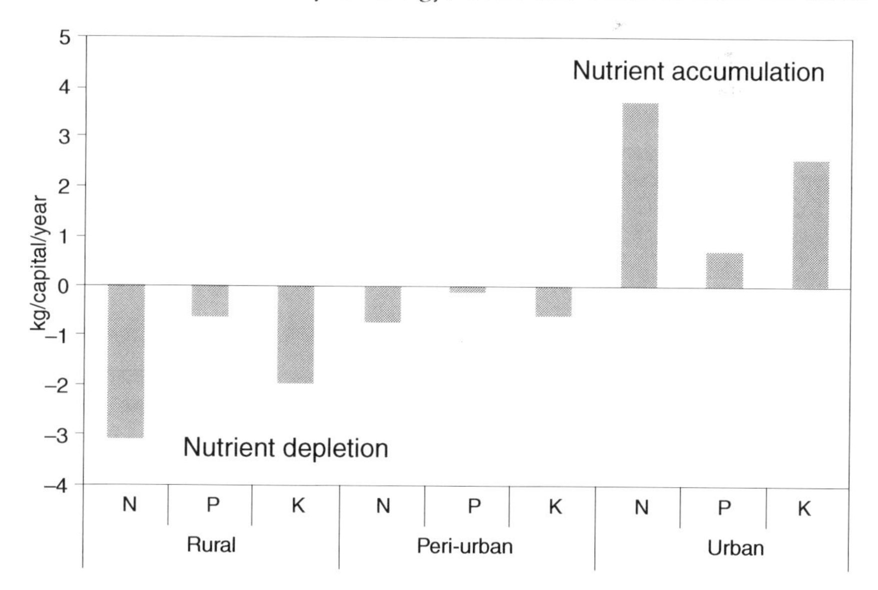

Figure 15.1 Urban nutrient accumulation and rural nutrient mining in kg/capita/year on average for four West African cities (Drechsel, 2014).

metabolism. Wastewater treatment across the four cities covers less than 10 percent of the households, which implies significant freshwater pollution and related increase of the urban grey water footprint.[1] A 'Material Flow Analysis' combining solid and liquid flows (Leitzinger 2001; Belevi, 2002, Erni *et al.* 2011) showed that between 70 and 80 percent of the N and P consumed in Kumasi pollute the urban environment, with ground and surface water receiving the largest share (Figure 15.2):

1. Groundwater and surface water receive about half of the total N and P losses, largely through leaking or overflowing septic tanks and latrines.
2. The result of this massive pollution is that the amount of N and P that leaves the city via waterways is over ten times more than the amount that enters the city in the same surface and subsurface streams and through precipitation.
3. About 22 percent of the N and 29 percent of the P enter the soil.
4. About 15 percent of the N is transferred to the atmosphere from households.
5. About 15 percent of both elements are sent with municipal waste to landfills.
6. Less than 5 percent of the N and P ends up in faecal sludge treatment plants.

Thus, only 20 percent of both nutrients is retained in the sanitation system, the rest contributes directly or indirectly to urban pollution (Erni *et al.*, 2011). The sad news is that Kumasi is in no regard an exception and reflects the common reality in sub-Saharan Africa.

From a technical perspective, part of this (environmental) load could be recycled – for example, through co-composting of faecal sludge and municipal solid waste. In a 'realistic' recycling scenario, which considered the actual waste collection capacity, the entire N and P demand of urban farming could be covered, as well as 18 percent of the N and 25 percent of the P needs of peri-urban agriculture within a 40 kilometre radius around Kumasi (Belevi, 2002). This transformation could easily replace any mineral fertilizer use in the whole Ashanti region around Kumasi, assuming a competitive price and product acceptance (Figure 15.3).

In the scenario of Figure 15.3 only the waste fluxes currently transported by truck to the landfill are considered. This solid waste originates from industry, market places, and from households and includes the biosolids from faecal sludge treatment ponds transported to the landfill.

Despite the gains for agriculture based on already collected waste the reduction of environmental pollution in this scenario was less than 20 percent, but it could be increased significantly with improved waste collection capacity. In particular, this concerns improved excreta management as the most significant N and P fluxes to soil, surface water and groundwater occur via faeces and urine transfer.

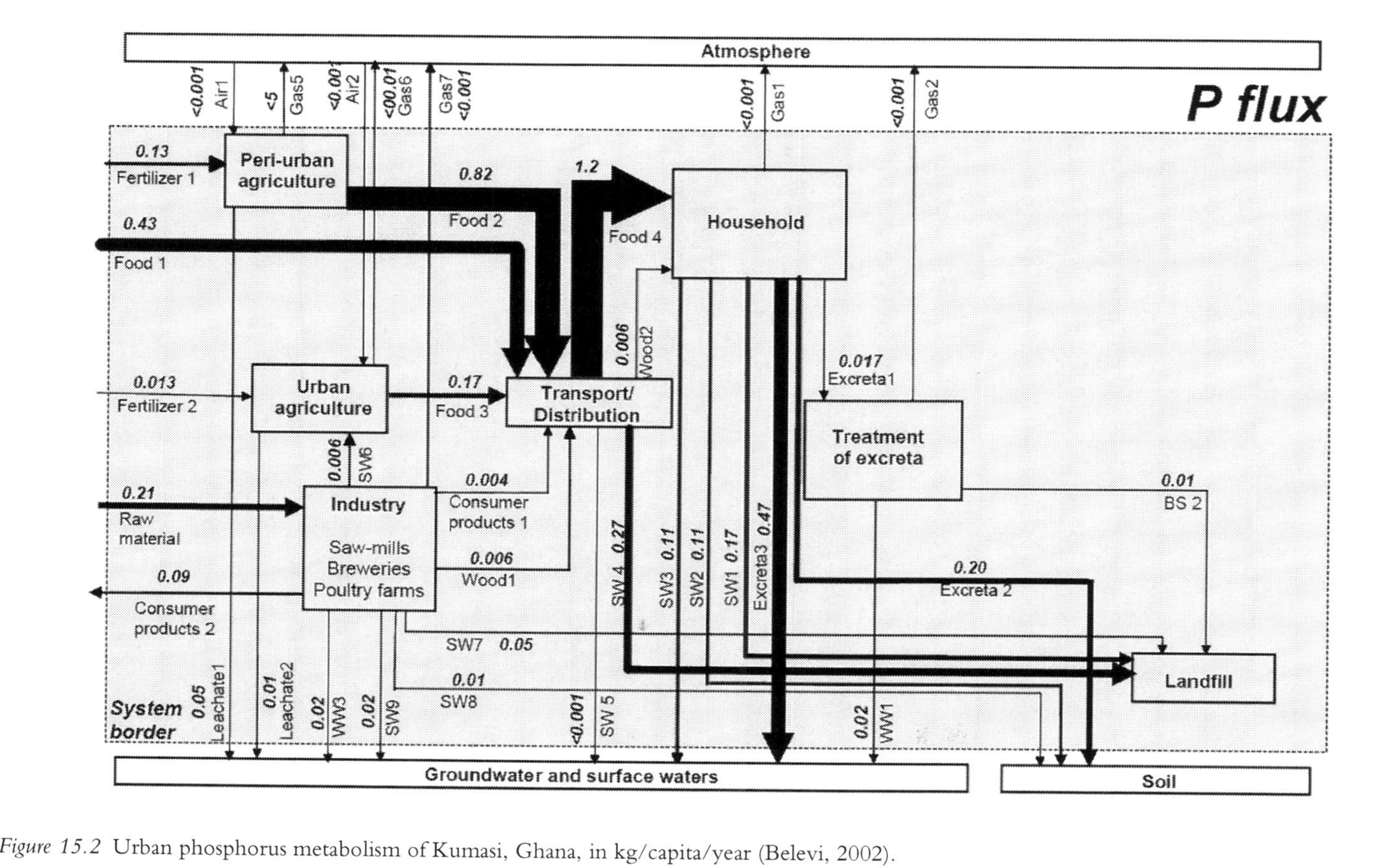

Figure 15.2 Urban phosphorus metabolism of Kumasi, Ghana, in kg/capita/year (Belevi, 2002).

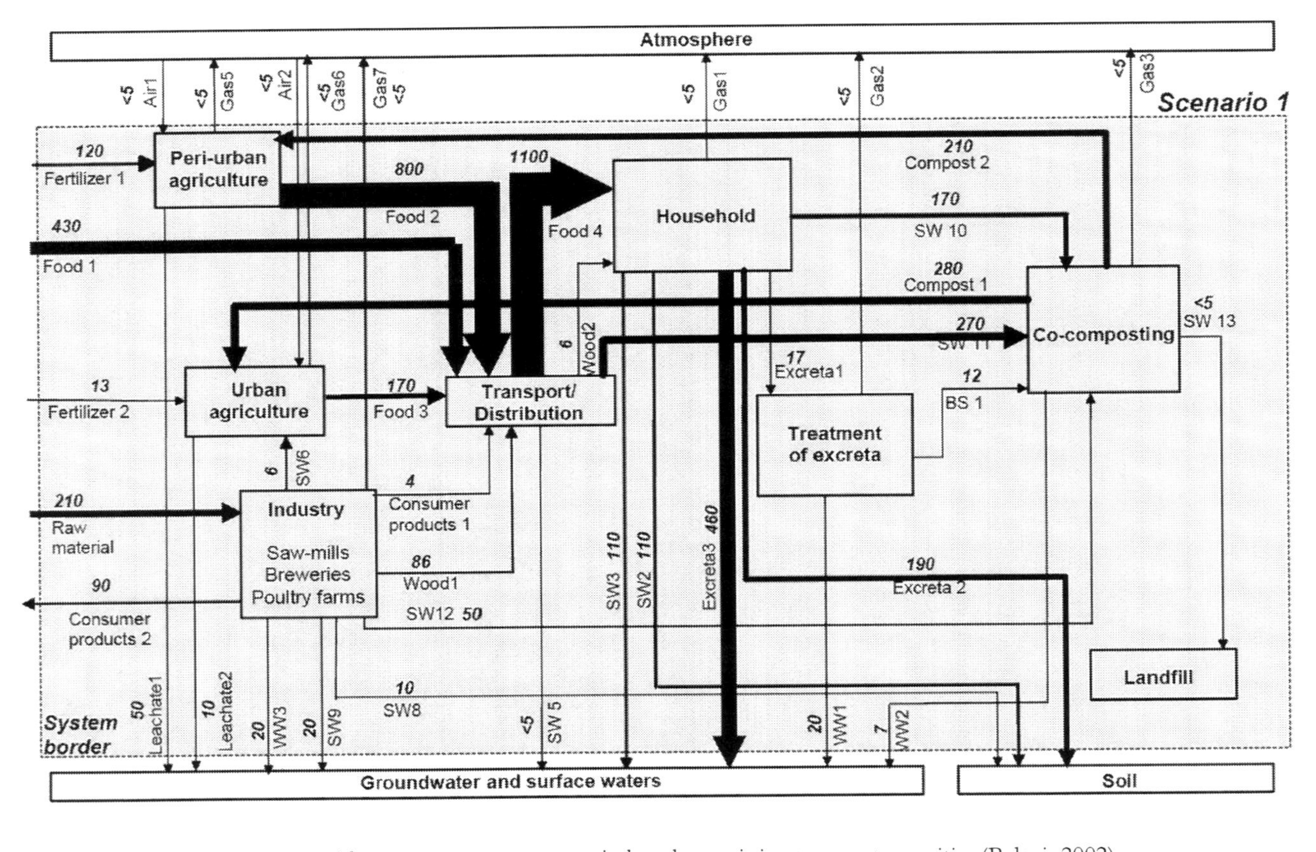

Figure 15.3 Phosphorus fluxes in t/year with resource recovery scenario based on existing transport capacities (Belevi, 2002).

Although there is no possibility to close a 'virtual' water loop, there are options for the urban metabolism to reduce its water footprint. A multi-purpose option is to improve resource recovery in terms of wastewater treatment and its productive reuse, especially in drier climates. This would not only bring a limited resource back into the production process but would also reduce the pollution of freshwater resources downstream of the cities (and the grey water footprint). Wastewater treatment in this context does not have to be energy-intensive or even conventional. In Accra, for example, irrigated urban agriculture recycles up to 10 percent of Accra's domestic wastewater, making the farming sector a larger informal 'treatment' service provider compared to that offered by public infrastructure (Lydecker and Drechsel, 2010).

Towards a circular economy

The current urban development model based on linear metabolism is no longer sustainable (Muradian *et al.*, 2012; Villarroel Walker *et al.*, 2014). Problems are mounting at both ends of the linear process, the rural production areas and the urban consumption areas (Figure 15.4). The target is a circular economy in which resources are reused again and again to achieve, in an ideal case, net zero waste and emission targets and promote recovery and reuse of water, energy and nutrients to underpin sustainable development within the rural–urban context.

Transforming urban society from a linear to a circular metabolism is being promoted in many developed countries. As a result, source separation is being introduced, cities and their environments are less polluted, resources are reused, and overall system resilience is increasing. This transformation remains a great challenge in low-income countries where public awareness about green values

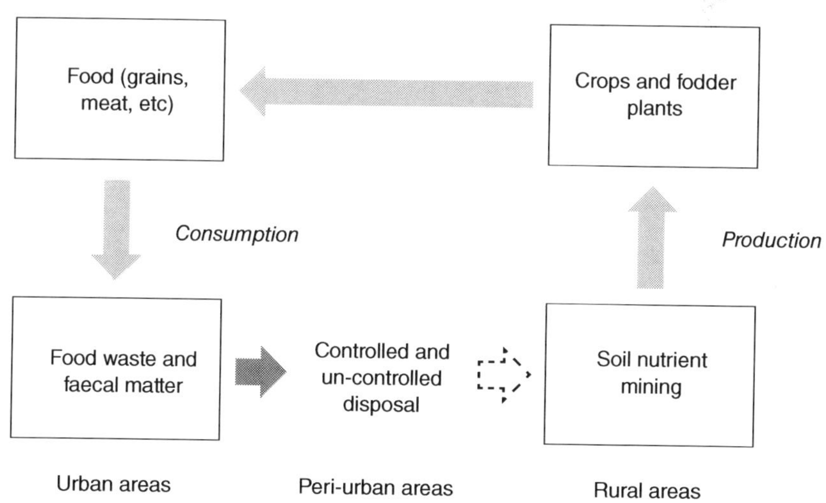

Figure 15.4 Linear resource flow from farm to waste with urban and peri-urban areas as vast nutrient sinks
(Source: Authors).

and opportunities is less developed and has a generally lower priority among many key stakeholders, and public budgets are too constrained to make the right investments. The likeliest best bet is to seek private capital and partnerships. While there are household- and community-based initiatives for resource recovery, success stories at scale are still limited to developed countries (Lazarova *et al.*, 2013), although exceptions are increasing where smaller and larger enterprises as well as public–private partnerships are presenting viable options for closing the loops (Otoo and Drechsel, 2015). A core principle of these businesses is full awareness of the Water–Food–Energy Nexus to take advantage of the economic opportunities urbanization offers. Many governments and private sector actors are beginning to realize the 'double value proposition', for example in water reuse: without reuse, wastewater treatment has a significant economic value in terms of environmental safety and public health, but almost no financial value. Reuse can add a range of new value propositions to the treatment proposition (Figure 15.5). An example is the lower energy costs of wastewater treatment compared to sourcing freshwater from long distances or the ocean (Voutchkov, 2010). Leveraging freshwater supply through the addition of recycled water can be a milestone of a circular metabolism as realized at scale, for example in Namibia (Rao *et al.*, 2015). Also, the recovery of energy from different types of waste or the recovery of P from wastewater treatment processes show how inter-sectorial approaches can push the circular agenda forward (Gebrezgabher *et al.*, 2015). In many situations, the direct revenues from selling treated wastewater are small, given that the freshwater prices are usually subsidized and the wastewater has to be sold even more cheaply. However, with the right business plan, several revenue streams can be combined. The simultaneous recovery of energy can allow wastewater treatment plants to become energy self-sufficient while supporting other reuse options which on their own would not be viable.

While energy might be a financial joker in the Nexus of resource recovery, it can also be a challenge. In particular, advanced wastewater treatment aiming at nutrient removal or potable water reuse requires significant amounts of energy while also targeting high-end markets (GWI, 2009; Lazarova *et al.*, 2012). However, especially in low-income countries energy might not only be expensive – it might also simply disappear for days or only come every other day, which will damage pumps and undermine aerobic processes; these are typical challenges which contributed in Ghana to the breakdown of many treatment facilities (Murray and Drechsel, 2011). A mismatch between imported treatment technologies and local requirements, possibilities and capacities has also been described, for example by Nhapi and Gijzen (2004) in Zimbabwe, and supports the call for low-cost applied technologies (Libhaber and Orozco-Jaramillo, 2013) with treatment levels cost-effectively matching the intended reuse or disposal (Murray and Buckley, 2010; Matos *et al.*, 2014). In areas where the bulk of the wastewater reuse is for restricted crop irrigation, advanced treatment may not be required and the use of appropriate technology can produce 'fit for purpose' water with much lower investment and

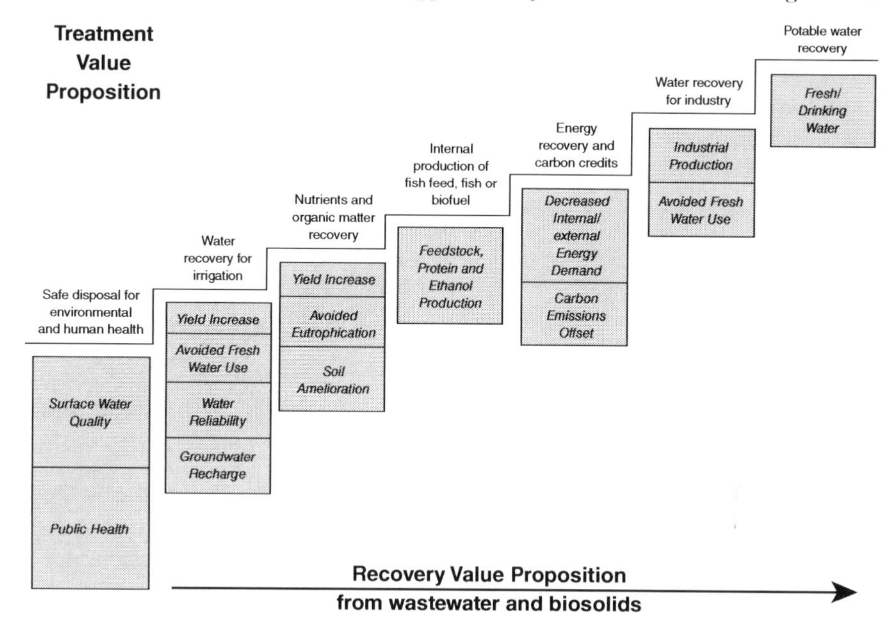

Figure 15.5 Value propositions related to water, nutrient and energy recovery from wastewater

(Source: Rao *et al.*, 2015).

maintenance costs, while entities requiring ultra-pure water can undertake their own treatment for high-end reuse. Thus, investments in sustainable resource recovery require careful judgements about whether technologies fit the local purpose and context.

Another common challenge, particularly in sub-Saharan Africa, is limitations in institutional capacities and finance. In particular, the lack of dedicated funds for infrastructure operation and maintenance can result in a run-to-failure trajectory starting with the commissioning of any new facility, as too often revenues from water tariffs or sanitation fees have to cover gaps in other municipal sectors (Murray and Drechsel, 2011). The example of the Drarga wastewater treatment plant near Agadir in Morocco demonstrates options for how to bypass this bottleneck. As in other locations, the municipality collects sewage fees to recover its operation and maintenance costs. In addition, the municipality designed the plant to generate additional revenue from the sale of: (i) treated wastewater to crop farmers; (ii) reed grass from the constructed wetland; (iii) sludge compost; and (iv) methane gas from energy recovery (Rao *et al.*, 2015). Although not all of these components have been implemented so far, a noteworthy innovation in this case is that all sales revenues and revenues from the water and sewage tariff and connection fees are deposited into a special account, independent of the main community account, to solely serve the wastewater treatment plant. This special arrangement is a response to common bottlenecks in the public financing of operations and maintenance, such as spare parts, which have contributed to the breakdown of about 70 percent of the wastewater treatment plants in the country (Choukr-Allah *et al.*, 2005). Other examples of business cases and models for resource recovery and reuse in low-income countries are presented by Otoo and Drechsel (2015).

Social challenges

Resource recovery and reuse is seen as a key process within a Nexus-based approach to support the sustainability of the urban metabolism within the rural-urban continuum. While related initiatives emerge across the developing world, they seldom reach a noteworthy scale or last for long. Given the availability of appropriate technology, the bottlenecks are probably in other sectors (Guest *et al.*, 2009). Next to a lack of economic, tax or social incentives for private sector engagement and business thinking in the sanitation sector, and common barriers such as a weak enabling environment from policies for institutions, public consultation and financing modes, a key bottleneck is the still overwhelming struggle against more basic waste collection problems. In more developed and risk-aware societies, public perceptions regarding reuse (Box 15.2) make stakeholder participation and trust building at the earliest stages of any reuse project crucial.

Box 15.2 Resistance to reuse

Toowoomba, in Queensland, Australia, is an often-cited case illustrating the strength of public opinion regarding wastewater use. A plan to turn wastewater into drinking water failed in Toowoomba at a referendum in 2006, although water scarcity in the community was severe, to the point that water use for gardening was completely prohibited in the 'Garden City'. With no major river nearby, the community water supply had to be pumped uphill. During several years of drought the 140,000 residents of Toowoomba and surrounding areas endured tough water restrictions. Local officials considered that the city had no choice but to treat and use parts of its wastewater for drinking water, and given the water crisis, they expected that the programme would be acceptable to the population. However, the proposal met with fierce opposition from the community. In 2006 the residents of Toowoomba voted strongly against treating and indirectly using 25 percent of the city's wastewater. They preferred to rely instead on water piped from Brisbane's Wivenhoe Dam, at a cost to ratepayers of nearly A$100 million more than the reuse programme would have cost. In San Diego, USA in the 1990s a similar social barrier prevented the use of already-installed treatment capacity (Hartley, 2006; Guest *et al.*, 2009; Drechsel *et al.*, 2015).

Path forward

Global energy and water cycles sustain life and economic activity on Earth, yet water, energy and food are entwined in complex ways – the Nexus, and the unequal distribution of energy and water resources across the planet, pose significant development challenges in matching local population demands, especially in urban agglomerations. Humanity has responded to these global metabolism challenges through interventions such as international and regional food trade and circular metabolism approaches supporting closed loop processes and ecosystem resilience. Reuse of water, nutrients and energy can optimize productivity in resource-scarce settings and support a move towards a circular economy model based on an integrated ecosystem approach. However, Guest *et al.* (2009) noted possible trade-offs within the Nexus through an overly heavy focus by the water industry on energy sustainability, potentially causing us to miss the major point of water sustainability.

Sustainable use of water, energy and nutrients holds a prominent position in the global ecosystem services and resilience debate. Population growth and rising incomes, combined with supply-side drivers such as climate change and the degradation of ecosystems, are creating acute conditions of water and nutrient scarcity which impact ecosystem health and productivity. Urbanization, as much as it is a challenge for resource allocation, also offers immense and so far hardly-used opportunities for resource recovery. Many of the related processes, like composting of organic waste, have very tangible benefits for waste management itself, even without taking advantage of any reuse market. There is a clear need to develop green solutions to urban sanitation challenges that deal with the Nexus between water, energy and nutrients and promote sustainable waste and sanitation service delivery, as well as valuing water, energy and nutrient recovery as the core of any urban resilience strategy. The emerging resource recovery and reuse portfolio offers business solutions across the agriculture–sanitation sectors, although still with quite different styles and rates of development (Otoo and Drechsel, 2015). The greatest challenge will be to support the transition towards a circular economy, especially in the exploding urban centres of low-income countries, based on clearly defined development targets coupled with sound business models and incentives which can attract private partners and capital.

Note

1 The additional environmental water requirements to dilute downstream contamination where wastewater treatment is poorly developed can easily multiply the normal urban water footprint.

References

Belevi, H. (2002) 'Material flow analysis as a strategic planning tool for regional waste water and solid waste management', Proceedings of the workshop "Globale Zukunft:

Kreislaufwirtschaftskonzepte im kommunalen Abwasser- und Fäkalienmanagement", Munich. Available online at: http://www2.giz.de/Dokumente/oe44/ecosan/en-material-flow-analysis-wastewater-and-waste-management-2002.pdf

Bird, J. (2014) 'Water-Food-Energy Nexus', in J. van der Bliek, P. McCornick and J. Clarke (Eds), *On target for people and planet: setting and achieving water-related sustainable development goals.* Colombo, Sri Lanka: International Water Management Institute (IWMI), 10–12,

Choukr-Allah, R., Thor, A. and Young, P.E. (2005) 'Domestic wastewater treatment and agricultural reuse in Drarga, Morocco', in A. Hamdy (Ed.), *The use of non-conventional water resources,* Bari: CIHEAM/EU DG Research (Options Méditerranéennes: Série A. Séminaires Méditerra éens; n. 66).

Craswell, E.T., Grote, U., Henao, J. and Vlek, P.L.G. (2004) *Nutrient flows in agricultural production and international trade: ecological and policy issues.* ZEF Discussion Papers on Development Policy No. 78. Bonn, Germany: ZEF (Center for Development Research, University of Bonn).

Drechsel, P. (2014) 'Who feeds the cities? A comparison of urban, peri-urban and rural food flows in Ghana', in B. Maheshwari, R. Purohit, H. Malano, V.P. Singh and P. Amerasinghe (Eds), *The security of water, food, energy and the liveability of cities: challenges and opportunities for peri-urban futures,* Dordrecht: Springer Water Science and Technology Library.

Drechsel, P. and Kunze, D. (Eds.) (2001) *Waste composting for urban and peri-urban agriculture – Closing the rural-urban nutrient cycle in sub-Saharan Africa.* Wallingford: IWMI/FAO/CABI.

Drechsel, P., Cofie, O. and Danso, G. (2010) 'Closing the rural-urban food and nutrient loops in West Africa: a reality check', *Urban Agriculture Magazine,* 23, 8–10. Available online at: http://www.ruaf.org/sites/default/files/UAM23%20west%20africa%20pag8-10.pdf

Drechsel, P., Graefe, S. and Fink, M. (2007) 'Rural-urban food, nutrient and virtual water flows in selected West African cities', Colombo, Sri Lanka: International Water Management Institute. Available online at: http://www.iwmi.cgiar.org/Publications/IWMI_Research_Reports/PDF/PUB115/RR115.pdf

Drechsel, P., Mahjoub, O. and Keraita, B. (2015) 'Social and cultural dimensions in wastewater use', in Drechsel, P. *et al.* (Eds), *Wastewater: economic asset in an urbanizing world.* New York: Springer.

Erni, M., Bader, H.-P., Drechsel, P., Scheidegger, R., Zurbrügg, C. and Kipfer, R. (2011) 'Urban water and nutrient flows in Kumasi, Ghana', *Urban Water Journal,* 8(3), 135–153.

Gebrezgabher, S., Rao, K., Hanjra, M.H. and Hernandez, F. (2015) 'Business models and economic approaches for recovering energy from wastewater and fecal sludge', in Drechsel, P. *et al.* (Eds), *Wastewater: economic asset in an urbanizing world.* New York: Springer.

GNA (2014) *'Pay Accra Compost Plant' – Zoomlion to Gov't.* GhanaWeb Regional News of Friday, 23 May 2014; available online at: www.ghanaweb.com/GhanaHomePage/NewsArchive/artikel.php?ID=310224

Grant, S.B., Saphores, J.-D., Feldman, D.L., Hamilton, A.J., Fletcher, T.D., Cook, P.L.M., Stewardson, M., Sanders, B.F., Levin, L.A., Ambrose, R.F. *et al.* (2012) 'Taking the "Waste" Out of "Wastewater" for human water security and ecosystem sustainability', *Science,* 337(6095), 681–686.

Guest, J.S., Skerlos, S.J., Barnard, J.L., Beck, M.B., Daigger, G.T., Hilger, H., Jackson, S.J., Karvazy, K., Kelly, L., Macpherson, L. *et al.* (2009) 'A new planning and design paradigm to achieve sustainable resource recovery from wastewater', *Environmental Science and Technology*, 43(16), 6126–6130.

GWI (2009) *Municipal water reuse markets (2010)*, Oxford, UK: Global Water Intelligence.

Hanjra, M.A., Drechsel, P., Mateo-Sagasta, J., Otoo, M. and Hernandez, F. (2015) 'Assessing the finance and economics of resource recovery and reuse solutions across scales', in Drechsel, P. *et al.* (Eds), *Wastewater: economic asset in an urbanizing world*, New York: Springer.

Hartley, T.W. (2006) 'Public perception and participation in water reuse', *Desalination*, 187(1–3), 115–126.

Lazarova, V., Asano, T., Bahri, A. and Anderson, J. (2013) *Milestones in water reuse: the best success stories*, London: IWA Publishing.

Lazarova, V., Choo, K.-H. and Cornel, P. (2012) 'Meeting the challenges of the water-energy nexus: The role of reuse and wastewater treatment', *Water21*, 14(2), 12–17.

Leitzinger, C. (2001) 'The potential of co-composting in Kumasi – quantification of the urban and peri-urban nutrient balance', in P. Drechsel and D. Kunze (Eds), *Waste composting for urban and peri-urban agriculture – Closing the rural-urban nutrient cycle in sub-Saharan Africa*, Wallingford: IWMI/FAO/CABI.

Libhaber, M. and Orozco-Jaramillo, A. (2013) 'Sustainable treatment of municipal wastewater', *Water 21 Magazine of the IWA*, October, 25–28.

Lydecker, M. and Drechsel, P. (2010) 'Urban agriculture and sanitation services in Accra, Ghana: the overlooked contribution', *International Journal of Agricultural Sustainability*, 8(1), 94–103.

Matos, C., Pereira, S., Amorim, E. V., Bentes, I. and Briga-Sá, A. (2014) 'Wastewater and greywater reuse on irrigation in centralized and decentralized systems: an integrated approach on water quality, energy consumption and CO_2 emissions', *Science of The Total Environment*, 493, 463–471.

Mensah, A. (2004) *Strategy for solid and liquid waste management in Kumasi*. Paper presented during the IWMISANDEC-KMA-KNUST Workshop on "Co-composting of solid and liquid waste for agriculture: Review of progress and next plans", January 27, 2004. KNUST, Kumasi, Ghana.

Mo, W. and Zhang, Q. (2013) 'Energy-nutrients-water nexus: integrated resource recovery in municipal wastewater treatment plants', *Journal of Environmental Management*, 127, 255–267.

Muradian, R., Walter, M. and Martinez-Alier, J. (2012) 'Hegemonic transitions and global shifts in social metabolism: implications for resource-rich countries. Introduction to the special section', *Global Environmental Change*, 22(3), 559–567.

Murray, A. and Buckley, C. (2010) 'Designing reuse-oriented sanitation infrastructure: the design for service planning approach', in P. Drechsel, C.A. Scott, L. Raschid-Sally, M. Redwood and A. Bahri (Eds), *Wastewater irrigation and health: assessing and mitigation risks in low-income countries*, UK: Earthscan-IDRC-IWMI, 303–318.

Murray, M. and Drechsel, P. (2011) 'Why do some wastewater treatment facilities work when the majority fail? Case study from the sanitation sector in Ghana', *Waterlines*, 30(2), 135–149.

Nhapi, I. and Gijzen, H.J. (2004) 'Wastewater management in Zimbabwe in the context of sustainability', *Water Policy*, 6, 501–517.

Otoo, M. and Drechsel, P. (2015) *Resource recovery from waste: business models for energy, nutrients and water reuse. A global catalogue*, US: Earthscan.

Otoo, M., Drechsel, P. and Hanjra, M.A. (2015) 'Business models and economic approaches for nutrient recovery from wastewater and fecal sludge', in Drechsel, P. *et al.* (Eds), *Wastewater: economic asset in an urbanizing world.* New York: Springer.

Otoo, M., Ryan, J.E.H. and Drechsel, P. (2012) 'Where there's muck there's brass: Waste as a resource and business opportunity', *Handshake (IFC)*, 1, 52–53.

Rao, K., Hanjra, M.H., Drechsel, P. and Danso, G. (2015) 'Business models and economic approaches supporting water reuse', in Drechsel, P. *et al.* (eds), *Wastewater: economic asset in an urbanizing world.* New York: Springer.

Rosemarin, A., de Bruijne, G. and Caldwell, I. (2009) 'The next inconvenient truth, Peak phosphorus', *The Broker*, 15, 6–9.

Scott, C.A., Zarazua, J.A. and Levine, G. (2000) *Urban wastewater reuse for crop production in the water-short Guanajuato River Basin, Mexico*, Research Report 41. Colombo: International Water Management Institute.

Ushijimaa, K., Funamizu, N., Takako Nabeshimab, Hijikata, N., Ito, R., Sou, M., Maïga, A.H. and Sintawardani, N. (2014) 'The Postmodern Sanitation – agro-sanitation business model as a new policy', *Water Policy*, uncorrected proof. DOI: 10.2166/wp.2014.093.

Van Rooijen, D.J., Turral, H., and Biggs, T.W. (2005) 'Sponge City: Water Balance of mega-city water use and wastewater use in Hyderabad, India', *Irrigation and Drainage*, 54, 81–91.

Villarroel Walker, R., Beck, M.B. and Hall, J.W. (2012) 'Water – and nutrient and energy – systems in urbanizing watersheds', *Frontiers of Environmental Science and Engineering in China*, 6(5), 596–611.

Villarroel Walker, R., Beck, M.B., Hall, J.W., Dawson, R.J. and Heidrich, O. (2014) 'The energy-water-food nexus: strategic analysis of technologies for transforming the urban metabolism', *Journal of Environmental Management*, 141, 104–115.

Vodounhessi, A. (2006) *Financial and institutional challenges to make faecal sludge management an integrated part of ecosan approach in West Africa.* Case study of Kumasi, Ghana. MSc thesis. UNESCO-IHE, Delft, the Netherlands.

Voutchkov, N. (2010) 'Seawater desalination: current trends and challenges', *Desalination*, 5(2), 4–7.

Part 6

Nexus corporate stewardship

How business is improving resource use

16 Building partnerships for resilience

David Norman and Stuart Orr

Why is this important to WWF and SABMiller?

This chapter is based on the experience of two organizations that have worked closely together to help define the role of business with regard to water, energy and food issues. It is not academically written, but comes more from a pragmatic sense of what we have learned along the way. It is therefore worth spelling out who we are and why we do this.

WWF is the world's largest independent conservation organization. Our mandate is to safeguard the natural world, conserving biodiversity and the vital benefits provided to humanity from natural systems. We see securing natural capital – the natural assets on which all our health and livelihoods depend – as an end in itself. However, we were also interested in the business case made by SABMiller and other companies in various sectors for approaching water, food and energy systems in a different way. To be clear, as an NGO we don't intend to push companies towards solving all their problems through eco-efficiency. Trying to tackle the major issues at hand through ever-greater efficiency won't enable us to meet the full range of challenges ahead. By moving into stewardship and collective action, we can help in creating guidelines around business risk, with multiple benefits for people and the natural world.

SABMiller is in the beer and soft drinks business, with 70,000 employees in more than 80 countries across six continents. Wherever we have operations, we require reliable supplies of good quality freshwater and agricultural products. Alongside this local dependence on natural resources, our business growth is also closely aligned with the transformation of the global economy, in particular through growth in emerging markets and the expansion in their middle classes. We see great opportunities in this transformation – both for our business and for the societies within which we operate. We see good stewardship of water, food and energy resources as the foundation of resilient social and economic development.

Global trends are entwined with great challenges for society and for all businesses (Waughray, 2011). While hundreds of millions of people have been lifted out of poverty, too many remain below the poverty line (Yoshida *et al.*, 2014). Threats from the erosion of natural capital are challenging the basis of development, and these threats are particularly acute for businesses that depend so directly on the natural world. If anything, the Nexus debate has emerged from this realization that development, national planning, regulations and policies have not kept pace with rapid social and economic change. In many ways policies designed to protect and share resources have been captured for narrow gain, translating risk on to other users and undermining the very natural capital that drives our economies (Hepworth, 2012).

Nexus interest from business also comes from a desire to engage with public decision-making bodies around the management of shared goods. The failure of their management (in water or energy, for example) is at the core of business exposure to risk, yet the involvement of the private sector in remedying this issue is controversial (Morrison *et al.*, 2010). Businesses need to acknowledge that much degradation has come about as the result of business operations taking advantage of unregulated resource use (Hepworth and Orr, 2013): public mistrust over their intent can be valid.

More than 70 percent of the world's freshwater use is in agriculture (Cosgrove and Rijsberman, 2014). Pressures deriving from water scarcity will only increase under new patterns of consumption driven by a population increase of three billion middle-class people by 2030, and a global population predicted to exceed nine billion soon afterwards (Godfray *et al.*, 2010). Much of the population growth and respective rise in middle-class growth will occur in countries that are already water-scarce. Further pressures arise from competing need for land, and from the imperative to decarbonize our energy systems. Climate change means that rainfall and water availability are likely to become more uncertain (Gosling and Arnell, 2013).

These trends matter for governments and businesses. They challenge the assumption that today's development strategies, which have delivered impressive poverty reduction and growth in prosperity over the past two decades, will continue to deliver into the future. They also challenge any model of business growth based on competition for natural resources, since all businesses and communities share the risk of resource scarcity – a risk that is not solved by seeking to control a higher share of a shrinking resource base (WWF, 2013).

Many of these pressures will be magnified by climate change, and there is already evidence to show that these shifts are putting greater stress on some river basins already compromised by the growing water needs of agriculture and industrialization (Gosling and Arnell, 2013). Some of the regions where food insecurity is most prevalent are the same regions where climate change is expected to pose the greatest challenges for agriculture. Ever-growing energy demands made by the world population exist in increasing tension with the urgent need to decarbonize and reduce the water intensity of our energy systems.

On one level, SABMiller experiences resource scarcity in terms of rising costs and the volatility in our raw material costs, but the threat is more fundamental than this. Without sufficient supplies of good quality water, our business could not exist. If rivers run dry, some of our businesses might have to shut down operations. The risk of water scarcity is a risk we share with the communities around us: these interdependencies demand collaborative responses. Yet identifying key stakeholders, policy gaps, trends and local issues of concern requires a daunting amount of tapping into local knowledge – knowledge which is often not easy for companies to attain. A great challenge going forward is to facilitate these types of engagements, lest we find companies falling back into positions of comfort based purely on internal efficiency gains that provide little connection with the regulatory or physical world outside.

The problem of unconnected responses

What kind of responses from companies and policy-makers will best ensure maintenance of prosperous societies and thriving natural systems in the face of those pressures? Organizations such as the CEO Water Mandate, part of the UN Global Compact, have been leading important debates on these issues over the last few years. They have developed guidelines on water accounting, collective action and public policy engagement, giving companies a much greater awareness around cooperative policy intervention (Morrison *et al.*, 2010).

The overarching response to these trends must be to find a strategic home for resilience within national development strategies and in business strategies. We cannot predict in detail what the consequences of population pressures, changes in consumption patterns and climate change will be. A resilience-driven approach acknowledges that there are many good reasons for building flexibility into our design and management of food, water and energy systems. For example, many of the policies that help to keep a river flowing under the immediate pressures from rising population and industrial uses will also build resilience, enabling that river system to continue to thrive in scenarios which encompass a wide range of possible climate change impacts.

National development strategies and business strategies need to be designed around this resilience, given the uncertainty in play: they need to work under many different scenarios representing possible outcomes of the trends described. Strategies need to build in capacity to absorb climate-related and population-driven shocks of many kinds, reducing the impacts of those shocks on people and on the natural systems on which we depend, thereby mitigating the likelihood, depth and frequency of those shocks.

Even in wealthy countries that can well afford to make this kind of investment in the nation's future, it is rare to find the uptake of genuinely helpful strategies. Our joint review of countries and states shows that, particularly at later stages of development, the inflexibility of institutions and the archaic state of infrastructure can be major risk factors undermining those countries' ability to respond to the climate-related and population-driven shocks being generated (SABMiller and WWF, 2014). The lessons for countries at a different stage – those with rapidly developing infrastructure and related institutions – could be critically important for their ability to sustain their development achievements into the future.

At the heart of the problem is a disconnect between the response and the problems identified. Many governments and businesses recognize the problems, but when the institutions that frame the responses are themselves broken into siloes, each may tackle a part of the problem in ways that conflict with solutions driven by another partial perspective. In some settings, the division between ministries within a government can epitomize the breakdown in communication. Finance ministries prioritize one set of criteria – for example, seeking rapid progress in solving power availability problems. Agriculture ministries may have another set of priorities, such as balancing rural livelihood development with the need to feed rapidly growing urban populations. Yet ministries responsible for water – which may be the limiting factor for any or all of these aspirations – are not necessarily made central to decision-making on strategy for energy and food systems in countries across the board.

The same can be true in companies. Technical responses to resource risks, such as driving up water efficiency within operations, are an essential part of the solution. In addition to offering cost savings, these responses can contribute to effective management of a local scarce resource, with the result that resilience for both the company and other local water users increases. Yet keeping too narrow a focus on technical solutions can distort priorities in situations where water risk is shared with other users, and where the vast majority of local water use occurs outside a company's operations. It is essential to complement operational efficiency with measures that tackle problems within a company's value chain and beyond (WWF, 2013). In order to improve water management beyond a company's value chain influence, partnerships must be created which reflect the fact that the majority of water use will not be limited to suppliers to the company, but will be inclusive of all water users. Partnerships need to reflect the broad band of stakeholders who share the risks and consequences of unsustainable water use, whatever its origins.

Opportunity missed

If problems are framed primarily in terms of future risk, this may undermine imperatives for government action. Within a company, quantifying value at risk can build a strong business case for action. SABMiller undertakes a country-by-country water risk assessment as the basis for informed choices about the most appropriate responses. For example, capital expenditure can be allocated with an expectation of specific return on that capital in terms of value protected or growth potential secured.

Immediate problems tend to monopolize government time, energy and resources. Power blackouts or food supply interruptions have immediate political consequences and governments may be under pressure to respond to these immediate crises in knee-jerk and sub-optimal ways (Orr and Cartwright, 2010). A different frame of reference is needed, which moves beyond the focus on future water and other natural resource pressures. Hundreds of millions of people around the world already experience the immediate consequences of past failures to focus on resilience (Adger, 2006). For them, water scarcity stands in the way of prosperity right now. They are farmers whose land could be more productive with efficient irrigation techniques. They are women who spend hours each day carrying water, instead of earning an independent income. They are children whose education stops when entire communities have to move in search of water in the dry season. They are small businesses with fewer customers because of the links between water scarcity and poverty. They are workers without jobs in places where companies are limiting their investment because of water insecurity.

Linking the resilience agenda with government responses to these immediate concerns is essential in order to build political space for action on the scale required. It can also help in developing a strong basis for local action by a mix of partners that are not used to working together.

Stewardship solutions

Through a growing body of work, managers have been seeking to define what stewardship is, and how it differs from internal business management (WWF, 2013). These debates mainly centre on water; although there is currently less engagement from the energy sector in terms of trade-offs and stewardship, certainly the food sector is increasingly focusing on this (World Economic Forum, 2010). If we continue to explore this notion through a water lens, then water stewardship serves to unite a wide set of stakeholders interested in external water management. WWF defines Water Stewardship for business as a progression of increased improvement of water use and a reduction in the water-related impacts of internal and value chain operations. More importantly, stewardship represents a commitment to the sustainable management of shared water resources in the public interest through collective action with other businesses, governments, NGOs and communities (WWF, 2013).

We recognize that business engagement in water management debates, and especially in public policy, provokes significant concerns from some NGOs and the public, including fears about business control of global resources. At the core of these concerns are two issues. Water is a highly complex public resource with multiple socially defined functions and values. Its effective management requires the continual reconciliation of trade-offs between private interests and collective well-being, not to mention fulfilment of a fundamental human right (Hepworth and Orr, 2013). Many people perceive an inherent conflict between a company's duty to shareholders and its ability to manage a wider set of social and environmental interests. Yet there is a growing group of companies that see the building of prosperous societies and the protection of thriving ecosystems as being so closely connected with the business's future that core business strategy must contribute to those social and environmental goals. Stewardship requires this insight: apparent trade-offs in the short term can become sources of alignment when a business's long-term interests are brought to the fore.

The social and environmental benefits arising from this insight can be immediate. The link between prosperity, water security and resilience is striking in some of the communities where both WWF and SABMiller operate.

In Neemrana, Rajasthan, for example, SABMiller's work with farmers on agricultural productivity is closely connected with solving problems of water scarcity. SABMiller India is enabling small-scale farmers to adopt rainwater harvesting, thereby increasing efficiency in food production through Community Watershed Management. Over the past three years, this project in Neemrana has helped farmers to increase their productivity by over a fifth, increase their water use efficiency by over a third, and at the same time enabled farmers to raise their disposable incomes by 21 percent.

The link between social, environmental and business outcomes suggests a central role in development policy for creating the conditions in which different actors collaborate to overcome the shared risks they face together. One example of such a collaborative approach has been the Water Futures partnership, of which SABMiller, WWF and GIZ were founding partners. It brings together businesses, donors, governments, farmers, communities and others who all face risks of increasing water scarcity in an area, and builds a local alliance to put in place practical measures to build resilience and secure the resource base on which they all depend. Policies that can catalyse and spread this kind of collaborative approach at scale are at the heart of a resilience-based development strategy.

The same logic, emerging from the close connection between risks to companies, water insecurity and untapped prosperity, leads SABMiller to look beyond local partnerships and work with others to engage in policy debates. For example, WWF and SABMiller jointly commissioned case-study-based research on the Nexus to inform policy development (SABMiller and WWF, 2014). While our two organizations have different starting points on this agenda, we see that even strong local partnerships are missing an important dimension of the way water, food and energy systems are managed. Governments, through regulations, incentives and national development plans, have a strong influence on the resilience of those systems in the face of climate and demographic change.

Examples of bringing policy responses and company responses together

The debate around notions of stewardship has also included questions of whether there should be an increased role for the private sector in development debates. From the WEF's attempts to raise interest and awareness through their work with McKinsey on water cost curves (Addams *et al.*, 2009), to the creation of a new set of public–private partnerships aimed specifically at delivering on development agendas, there are legitimate questions around public acceptance of a private sector agenda which is being given increased consideration.

Nevertheless, groups such as the International Council for Mining and Metals (ICMM) have developed progressive water strategies to help their member companies (over twenty of the largest mining companies in the world) to engage in 'catchment approaches' in order to deal with long-term resource risk (ICMM, 2014).

Collaborations at a national and international level are being developed to implement practical initiatives to address the Nexus issues being addressed. The Consumer Goods Forum is a collection of over four hundred retailers and manufacturers which has made commitments to help achieve zero net deforestation by 2020, as well as phasing out harmful refrigerants by 2015. The World Business Council for Sustainable Development (WBCSD) continues to lead its members into greater awareness of resource valuation and its implications (WBCSD, 2012). The CEO Water mandate and the New Vision for Agriculture (NVA) continue to push the boundaries on company participation on water and food debates at national and river basin scales, and NGOs like WWF are now positioning themselves to convene these landscape-scale strategies to work with public and private actors around ideas of collective action.

Conclusion

The Nexus debate has followed closely on the heels of carbon and water challenges and in many ways offers a more complete expression of what has been learned. The ultimate risk for companies arises not so much from the incremental impacts they may have on resources, but from the growing challenges of the global management and reconciliation of societal and business needs related to the production of water, food and energy. Their collective management, and our ability to design a linked set of business and societal systems of resilience, will remain the paramount challenges as we move towards a warmer and more crowded world.

References

Addams, L., Boccaletti, G., Kerlin, M. and Stuchtey, M. (2009) *Charting our water future: economic frameworks to inform decision-making*, 2030 Water Resources Group: McKinsey & Company. Available online at: www.2030waterresourcesgroup.com/water_full/Charting_Our_Water_Future_Final.pdf

Adger, W.N. (2006) 'Vulnerability', *Global Environmental Change*, 16(3), 268–281.

Cosgrove, W.J. and Rijsberman, F.R. (2014) *World water vision: making water everybody's business*, London: Routledge.

Godfray, H.C.J., Beddington, J.R., Crute, I.R., Haddad, L., Lawrence, D., Muir, J.F. and Toulmin, C. (2010) 'Food security: the challenge of feeding 9 billion people', *Science*, 327(5967), 812–818.

Gosling, S.N. and Arnell, N.W. (2013) 'A global assessment of the impact of climate change on water scarcity', *Climatic Change*, 1–15.

Hepworth, N.D. (2012) 'Open for business or opening Pandora's Box? A constructive critique of corporate engagement in water policy: an introduction', *Water Alternatives*, 5(3), 543–562.

Hepworth, N. and Orr, S. (2013) 'Corporate Water Stewardship: new paradigms in private sector water engagement', in B.A. Lankford, K. Bakker, M. Zeitoun and D. Conway (Eds), *Water security: principles, perspectives and practices*, London: Earthscan Publications.

ICMM (2014) *Water stewardship framework*, London, ICMM. Available online at: www.icmm.com/document/7024

Morrison, J., Orr, S., Schulte, P., Hepworth, N., Christian-Smith, X. and Pegram, G., (2010) *Guide to responsible business engagement with water policy*, UN Global Compact, Oakland: Pacific Institute/The CEO Water Mandate.

Orr, S. and Cartwright, A. (2010) 'Water scarcity risks: Experience of the private sector', in L. Martinez-Cortina, A. Garrido and E. Lopez-Gunn (Eds), *Re-thinking Water and Food Security*, London: CRC Press.

SABMiller and WWF (2009) *Water footprinting: identifying and addressing water risks in the value chain*, Surrey, England: SABMiller and WWF-UK.

SABMiller and WWF (2014) *The Water-Food-Energy Nexus: insights into resilient development*. Available online at: www.sabmiller.com/docs/default-source/investor-documents/reports/2014/sustainability-reports/water-food-energy-nexus-2014.pdf

Waughray, D. (Ed.), (2011) *Water security: the water-food-energy-climate nexus*. Washington, DC: Island Press.

WBCSD (2012) *Water valuation; Building the business case*, Geneva: WBCSD Water.

World Economic Forum (2010) *Realizing a new vision for agriculture: a roadmap for stakeholders*, Geneva: WEF.

WWF (2013) *Water Stewardship – a WWF perspective*. Switzerland: Gland.

Yoshida, N., Uematsu, H. and Sobrado, C. (2014) *Is extreme poverty going to end? An analytical framework to evaluate progress in ending extreme poverty*, Washington, DC: World Bank. Available online at: https://openknowledge.worldbank.org/handle/10986/16811

17 Capital markets at the Nexus of sustainable development

Steve Waygood

Introduction

For generations policy makers have sought to align the interests of the financial markets and society. Nowhere is this tension more keenly and persistently felt than in the relentlessness of the capital markets in allocating capital to short-term, unsustainable uses, in combination with policy makers' need to plan for the long term and tackle a range of environmental and social issues such as the nexus of issues surrounding poverty, climate change, fresh water and human rights.

The purpose of this chapter is to provide the people involved in policy making with an overview of how the equity market is structured, as well as some specific suggestions as to how they can move the capital markets to a more sustainable basis.

In particular, this chapter is focused on the creation of policy at the governmental and intergovernmental levels, as well as think tanks and non-governmental organizations who share our desire to build markets on a sustainable foundation. Our immediate focus is those involved in the Implementation of Transforming Our World: The 2030 Agenda for Sustainable Development, which replaced the Millennium Development Goals at the end of 2015, but the ideas will remain pertinent beyond this. To be even more specific, I see the emergence of the water, energy, food and climate Nexus as the centre of the challenge for many companies.

Adopting the conventional definition of sustainable development for the finance sector, we are seeking to promote capital markets that finance development that meets the needs of the present, without compromising the ability of future generations to meet their own needs. The use of the bundling of similar and interconnected factors, as seen within the Nexus, represents the inclusion of sustainable accounting and development.

As investors and insurers, our commercial concern is the contemporary threat to financial stability and long-term economic growth that originates from the unsustainable use of natural and social capital. It is clear to us that one of the underlying causes of the financial crisis in 2008 was the incentive structure throughout the markets. This focused too many market participants on short-term profits. They looked only so far as the next quarter's earnings, at the expense of paying attention to the longer-term fault lines that were emerging. Compounding the problem, much of the information available to investors – particularly on the environmental and social impact of a company, on financial structuring and business practices – was itself short-term and inadequate. This lack of information eventually negatively affected the entire market.

This comes at a time when predictions from the Stockholm Environment Institute (SEI) show that we will need to find something like 30 to 40 percent additional food in the coming fifteen years, and a similar amount of energy. Underlying this will be a 30 percent shortfall in water availability over demand. As investors, we need to know how companies are addressing the Nexus as it will have a huge impact on profits and the sustainability of companies.

A stable macro-economic environment is also important for the maintenance of financial stability. Prudent monetary and fiscal policy is important in avoiding speculative bubbles. So are economic reforms that facilitate the creation of an economic environment where real interest rates can be low and stable, and growth can be achieved, which is key in maintaining long-term stability in the sector.

The relevance of the capital markets to sustainable development policy makers

It is perhaps obvious to say that electronic flows of money around capital markets have no tangible impact on our physical environment, nor on our society. However, tangible impacts certainly do arise when the capital is spent on the production or consumption of goods and services – for example, the environmental costs and social benefits that arise from the development of infrastructure projects by companies. Importantly, impacts also arise when capital is required for a development project but is not forthcoming.

Policy makers have traditionally tended to focus on the flow of aid, or official development assistance. This is important, and the US$130bn of official development assistance matters. However, how the hundreds of trillions of assets under management within the global capital markets are allocated matters far more.

The capital markets are of relevance to policy makers for three distinct but related reasons:

1. As a way of *raising capital* to enhance government spending on Nexus-centred, sustainable development projects;
2. As a target for systemic change to integrate sustainability at each stage as the *financial influence* of the capital market via corporate access to capital can enhance or undermine long-term sustainable development goals; and
3. As an *ownership mechanism* for influencing corporate practices that policy makers can seek to harness in order to improve the sustainability practices of existing listed companies.

The discussion below expands on these areas. To achieve sustainable markets, policy makers need to both change the pricing signals within the market and improve the readiness of the supply chain of capital to integrate sustainability issues. This involves moving all participants towards a longer-term perspective when investing and exerting their influence as company owners. This requires an overview of the role of institutions at each stage of the capital supply chain.

A brief introduction to the equity market

To help those involved in policy making to picture the system, we have produced Figure 17.1 below. This depicts the relationship between the financial institutions that operate the market between the demand for and supply of capital. The different roles of the financial institutions are important as each role reflects the nature of the influence. We will also use this systems map as a mechanism for analysing equity markets and making recommendations as to how sustainability can be integrated into the system.

Figure 17.1 is a simplified model of the equity capital market and is intended to demonstrate how the various capital market institutions relate to each other. Put simply, money flows from the individuals on the right-hand side into companies on the left-hand side, which put the capital to work in different ways to generate a return on investment for their shareholders. These individuals may invest alongside others as institutional investors – such as pension schemes, insurance companies, or sovereign wealth funds – or as individual investors through retail financial advisors.

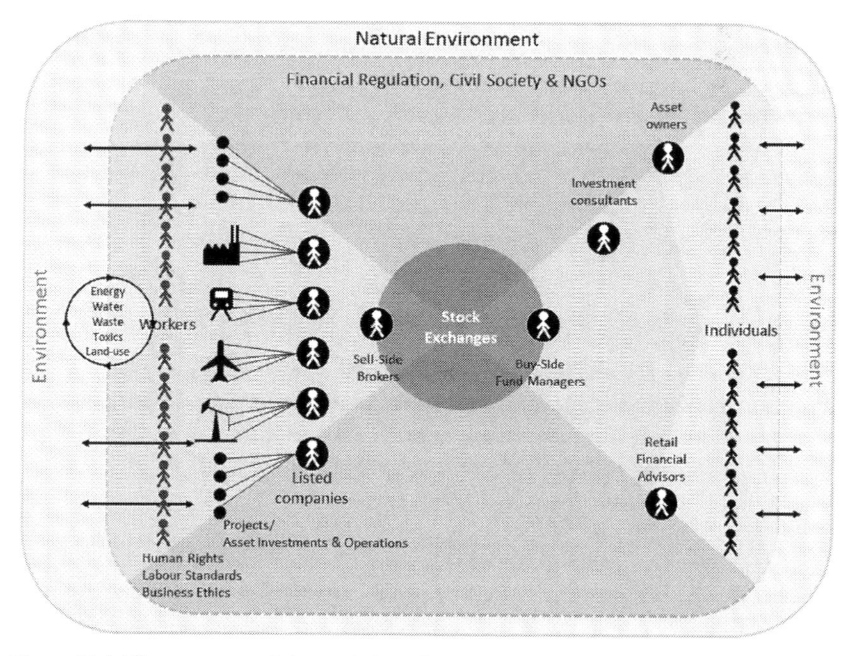

Figure 17.1 The structure of the capital market.

Corporate debt is also traded in a similar way, although it tends to be conducted between fund managers and brokers without the use of stock exchanges.

Of course, while company workers are highlighted as working for companies, they are also in many cases the providers of capital. So, in this sense the flow of capital can be thought of as circular.

Taking a look at this system in more detail, the stock exchange intermediaries and institutions include, among others, stockbrokers, fund managers, issuing houses, merchant banks (now more commonly called investment banks) and, as general buyers and sellers of securities, the central bank, commercial banks, pension funds, insurance companies, unit trusts, mutual funds, investment trusts, open-ended investment companies and company treasuries. Here they collectively represent the main types of capital market financial institutions in question.

It is necessary to express how the main types of financial institutions relate to each other so that we can be clear about the chain of influence that makes the capital market of interest to policy makers. The following section describes the capital market institutions that facilitate the flow of capital from investors (which supply the capital) to companies (which demand the capital).

Figure 17.1 shows that in the UK, the supply of equity capital originates from two main areas:

1. Individual Investors – individuals, either as scheme beneficiaries or directly as 'retail' investors, purchasing stocks and shares from an investment broker, or investing in pooled schemes such as open-ended investment companies (OEICs), *ociedad de inversión de capital fijo* or *société d'investissement à capital fixe* (SICAVs), unit trusts and investment trusts managed by fund managers; and
2. Institutional Investors – such as company and local authority pension funds, insurance companies, investment trusts, charities and organizations operating unit trusts and investment trusts.

The demand for equity capital comes from companies (PLCs) listed on stock exchanges. Globally, the value of all companies listed on stock exchanges is over US$50 trillion. These PLCs use the services of investment banks to underwrite the new issues of their shares (it should be noted that in many developed markets the use of equity finance by listed companies is becoming a less significant source of capital for companies than finance via corporate debt).

Investment banks also have a role in facilitating mergers, acquisitions and new placements on the exchange. Furthermore, many investment banks include sell-side broker operations (institutions that sell equities to investors for a percentage commission) that act as intermediary agents between companies and investors, maintain markets for previously issued securities and offer advisory services to fund managers. This last advisory role is one service that renders sell-side brokers important to policy makers who work on Nexus issues. Fund managers place considerable reliance on the views of these analysts, with the consensus in their forecasts being a closely monitored factor by many analysts. Where the views of the most influential brokers change, markets also tend to move: consequently, brokers' views on Nexus practices and sustainable development will be influential.

Buy-side fund management houses (institutions that buy and hold securities in the expectation of a return on investment) buy and sell their equities via sell-side brokers. They may also use their advisory services. It is the job of the individual fund manager to make individual portfolio investment decisions in accordance with the stated aims of the investment fund, and may also employ their own internal analysts. The client's aims are set out by the asset owners in the investment mandate, which is also known as the Investment Management Agreement.

Similar to retail investors seeking the advice of independent financial advisors (IFAs), institutional investors place considerable authority in the views of investment consultants who advise as to which fund manager has the most robust investment process and can meet the investment needs of the investment scheme. This is partly for legal reasons in that a trustee has the responsibility to represent pension fund beneficiaries, and partly for practical reasons in that many trustees do not have the professional skills to assess the investment processes of fund managers. This is particularly the case in the UK, the United States and Canada. Therefore, being able to articulate a robust investment process that impresses investment consultants is of central importance to fund managers. This is because they must be able to convince the investment consultants that they have the people, the investment philosophy and the investment process that should deliver consistent performance in order to win business. Consequently, fund managers spend a considerable amount of time and effort on the areas that investment consultants rate as important aspects of a good process.

Investment consultants are highly relevant to policy makers because they significantly influence institutional investors' choices of fund manager. Consequently, if investment consultants indicate that they believe that something is important, this sends a powerful market signal to fund managers, who are more likely to invest more resources in this area as a result.

Capital market sustainability failure

The primary failure of the capital markets in relation to Nexus issues is one of the misallocation of capital. This in turn is a result of the failure of global governments to properly internalize environmental and social costs into companies' profit and loss statements (through, for example, fiscal measures, standards, regulation and market mechanisms). As a consequence, the capital markets do not incorporate companies' full social and environmental costs. Indeed, until these market failures are corrected through government intervention of some kind, it would be irrational for investors to incorporate such costs since they do not affect financial figures and appear on the balance sheet or – therefore – affect companies' profitability or earnings per share. This means that the corporate cost of capital does not reflect the sustainability of the firm. The consequences of this are that unsustainable companies have a lower cost of capital than they should, and vice versa.

This is a critical point to understand, as it is central to how the power of the market can be harnessed by policy makers to promote sustainable development. Put simply, if it costs Company A more to invest in its own growth than it costs Company B, then all other things being equal, Company B will prosper. As mentioned before, this is why the aggregate investment decisions of analysts and fund managers are so powerful: the pricing signals we send to corporate managers influence the terms on which they raise capital. These terms can be central to achieving a competitive advantage and influence the ease with which companies go about their businesses. It is in this sense that the market can be seen as failing to motivate corporate practices that promote an integration of Nexus issues into companies' core business.

As the World Business Council for Sustainable Development (WBCSD) has noted: "*Financial markets are key in the pursuit of sustainable development because they hold the scorecard, allocate and price capital, and provide risk coverage and price risks*" (WBCSD, 2003, p6). If financial markets do not understand and reward sustainable behaviour, progress in developing more sustainable business practices will be slow. This requires the kind of system interventions by governments that change a company's cost of capital and harness the influence of company owners.

One of the reasons for emphasizing this market failure point is that government representatives can often direct market participants to embed sustainability issues into their valuation work. However, until the policy makers themselves correct the market failures, this directive will amount to little more than a rhetorical flourish.

Through the integration of multiple sustainability fields and integrating Nexus principles (see Chapter 18), sustainability can be accounted for as a unit, which can be integrated into current accounting methodologies more comprehensively. It is important that the push to include more environmental and Nexus principles in this process should be set forth in a clear and comprehensive manner, because the failure to do so could encourage furthering the status quo or – possibly just as harmfully – lead companies to account for either the wrong or incomplete factors.

Raising capital to fund government spending on sustainable development projects

Capital markets provide funding for public goods via a range of mechanisms, for example: sovereign debt, which is issued by countries; multilateral development bonds, which are issued by development banks; and municipal bonds, which are issued by local authorities.

Sovereign debt is one of the asset classes best predisposed to provide the capital to finance an approach to the challenge that the Nexus gives us – and building resilience is part of this. The speed and scale of the growth in sovereign debt that was issued to underpin the global financial system during the financial crisis demonstrates that it is possible to secure financing at the speed and scale implied here. The key question is the existence of political will to direct that finance towards the Nexus.

The issuance of new sovereign debt during the financial crisis has tested the credit ratings of a significant number of countries, stretching them to the point where it is unlikely that they will have the economic capability, let alone the political will, to make new issuances at the scale required here. Of course, the multilateral development banks (MDBs) would also have a role to play in issuing new debt that could finance these sustainable development plans. Indeed, there has already been some work in this area from the World Bank. However, this work has not covered the scale required, and many green bond issuances have been at a premium to the main MDB debt. This renders the products suitable largely only for the niche green investment community and not the mainstream investment industry, severely limiting the scale of the capital that can be raised.

In view of the scale of financing required – and the need for governments to also underwrite the MDBs – it is possible that there will come a point where the MDBs will also be unable to issue new debt on this scale without affecting their credit rating, and those of the countries supporting them. This will make debt more expensive as well as far less palatable politically. Therefore, post the financial crisis the two important mechanisms for raising this capital may be constrained in their ability to do so.

In terms of specific sustainable finance policy recommendations on *raising capital*, if policy makers are to raise this money in an efficient, effective and sustainable manner, we believe that the international community should develop a series of well-considered national capital-raising plans that are internationally coordinated at the UN and World Bank level, and that include a view on the money that can be raised via infrastructure investment, project finance, corporate debt, foreign direct investment, equity investment, and sovereign and MDB debt. Such cooperation would further the base of Nexus connections addressing the creation of more resilient communities, helping to push the principles to a wider, more secure and more influential audience.

The capital markets' *financial influence* over corporate sustainable development

The capital market's financial influence over corporate sustainable development originates from the buying and selling of equity shares and debt on the capital market. As noted earlier, this trading activity influences the cost of capital for listed companies, which is the price the company has to pay to raise capital to finance its business. This is generally referred to as a Weighted Average Cost of Capital (WACC). Mathematically, the WACC equation is the summation of the cost of each capital component multiplied by its proportional weight. Broadly speaking, a company's assets are financed by either debt or equity.

WACC is the average of the costs of these sources of financing, each of which is weighted by its respective use in the given situation. By taking a weighted average, we can see how much interest the company has to pay for every dollar it finances. Put simply, the more a company has to pay for capital, the less it can raise. This limits the extent of its activity. In addition, the financial value of the shares influences a director's remuneration and the degree to which the company is perceived as a candidate for takeover.

There is mounting evidence that companies with strong sustainability performance deliver improved long-term returns. For example, a July 2013 Harvard Business School study found that "High Sustainability" companies significantly outperform their peers over the long term, in both stock market and accounting terms (Eccles, Ioannou and Serafeim, 2013). This finding is consistent with a Deutsche Bank study providing a meta-analysis of over one hundred academic studies (Fulton, Kahn and Sharples, 2012).

So if sustainability performance matters to companies and their shareholders, what is the problem with the capital market?

The key sustainability problem with the existing capital markets is that the cost of capital for companies is not *sufficiently* influenced by how sustainable the company is. In other words, sustainability issues and particularly issues relating to the Nexus do not matter enough to ensure that performance is sustainable in the short term. We believe this is the case for two related reasons: market inefficiency and market failure.

Market inefficiency is used here to refer to the situation where it pays for companies to do the right thing and be sustainable, but markets neither recognize nor reward this behaviour until the company delivers the results within their accounts. In other words, while companies plan to be sustainable, investors do not proactively see the reflected impact in a business sense, and do not recognize how their initial investment has an impact towards lowering a company's short-term cost of capital, until the benefits are obvious to all. This time lag can punish more sustainable companies via a higher cost of capital until the benefits of their behaviour appear in their accounts.

On the other hand, market failure refers to the situation where it pays companies in the long term to do the wrong thing and be unsustainable. In other words, a market failure is where the externalities associated with unsustainable business practices do not hit the company's profit and loss (P&L) statement at all. As mentioned earlier, this is largely because global governments have not taken corrective action to internalize the costs onto corporate balance sheets.

The difference between capital market inefficiency and capital market failure is that the former is a failure of the predictive power of investors, whereas the latter is a failure of the governments to create a market price mechanism that ensures that companies have to pay the cost of their externalities.

A cause of market inefficiencies: misaligned incentives

One of the main sources of market inefficiency is an excessively short-term view among market participants who are more concerned about short-term costs or benefits of an initiative than the long-term costs or benefits arising from it. The short-term argument rests on the capital market being too near-sighted in the way it evaluates companies.

One root cause is that fund management organizations are evaluated by their clients – for example, pension funds – based on criteria that are themselves too short-term. Such evaluation motivates short-term investment behaviour on the part of fund managers that is more akin to speculation than to genuine ownership. Fund managers are subject to a legal fiduciary duty to obtain the best risk-adjusted financial returns for their clients, and this is often evaluated on the basis of very short-term, even daily results. In an ideal world, their interest would be in the long term, but the structure of the market pushes them into maximizing short-term returns.

As previously outlined, this maximization of short-term results is a long-term problem for the economy as a whole: if the capital market does not sufficiently factor in long-term capital investment returns, then it undermines long-term investment decision making by company directors and leads them to allocate insufficient capital to investing in the long-term health of companies overall. While a lack of focus on the long-term financial health of a company is a general problem, short-termism is also a particular problem for sustainable development: it systematically erodes incentives for company directors to invest in a sustainable business.

A Tomorrow's Company report commissioned by Aviva Investors confirmed this thinking and found that potential conflicts of interest in the capital market supply chain shown in Figure 17.1 above include:

1. Pension fund trustees: the close and frequent monitoring of fund management performance by trustees can result in fund managers feeling pressured to maintain high levels of short-term performance relative to the benchmark to retain funds; 66 percent of pension funds formally review fund manager performance every quarter (92 percent annually or less), despite 62 percent of them claiming that the key investment period for trustees appearing is longer than a rolling or calendar year (Source: "NAPF/IMA Short-Termism Study Report", MORI, 2004). This can create incentives that affect fund managers' approach to risk taking.
2. Investment consultants: the degree to which investment consultants take into account factors relating to the long-term sustainability of companies is dependent on: (i) the degree to which pension fund trustees wish to take them into account; (ii) the cost of maintaining dedicated research teams; and (iii) the availability of good long-term comparable data. Investment consultants also tend to charge a fixed

hourly rate and therefore have an incentive to be active in order to maximize their income. They therefore offer an increasingly wide range of services that they encourage trustees to use. There is an opportunity to generate substantial income through the fund manager selection process, so consultants may be incentivized to encourage fund manager turnover.

3 Brokers: the remuneration of brokers is directly linked to trading volumes. As a result, they have a powerful incentive to encourage market activity. Even when sell-side analysts are aware of corporate governance or sustainability concerns, these analysts do not report this in their reports to buy-side analysts for fear of losing access to those boards.
4 Stock exchanges: nearly half of all exchanges are companies listed on their own exchange, and they are therefore subject to shareholder pressure to maximize returns. The largest sources of revenue for demutualized, for-profit stock exchanges are reliant on market activity. This results in an incentive for exchanges to create inducements for trading activity.

In order to change this, simple measures could be implemented to align these incentives. For example, to expand on the specific example regarding investment consultants: the United Nations Environment Programme Finance Initiative's Asset Management Working Group (UNEP FI AMWG) has called for actions that they believe will magnify the extent to which responsible investment is demanded by the capital markets. They proposed: "Global capital market policymakers should make it clear that advisors to institutional investors have a duty to proactively raise environmental, social and corporate governance (ESG) issues within the advice that they provide, and that a responsible investment option should be the default position. Furthermore, policymakers should ensure prudential regulatory frameworks that enable greater transparency and disclosure from institutional investors and their agents on the integration of ESG issues into their investment process, as well as from companies on their performance on ESG issues" (UNEP, 2009).

The report also proposes clauses on responsible investment that should be written into fund management contracts, and that fund managers' performance should be based on an evaluation of their long-term ability to beat benchmarks. Moreover, because of the potential for unsustainable development to harm the absolute value of long-term investment portfolios, the report also proposed that investment consultant and fund manager clients should be able to sue for negligence if these issues are not properly considered.

More generally, we believe that various market participants should be incentivized to change their behaviour in order to achieve a culture shift in financial institutions. For example, we support the abolition of quarterly reporting, and similarly we believe that fund manager performance should be reviewed over longer time horizons instead of the typical quarterly cycle. Such evaluation based on short-term returns undermines long-term sustainable development; for the same reason, excessive reliance on measuring performance relative to a market index should be reduced.

In order to align incentives over the long term (as highlighted by Tomorrow's Company), pension funds should be required to integrate voting and engagement policies into the investment process; shareowner activism should be given more weight in the selection and retention of fund managers and other matters. In addition, all advisors to institutional investors should have a duty to proactively raise ESG issues and encourage adherence to the Stewardship Code, and fund management contracts and fund managers' performance should include an evaluation of long-term ability to beat benchmarks. Investment consultants' fee structures should not reward them for moving clients between fund managers, and within companies the implementation of strong cultural norms should be supported by independent whistle-blowing mechanisms, overseen by professional bodies who offer the whistle-blower appropriate protection.

A cause of market inefficiencies: A lack of information from companies to the providers of capital

The ability to assess a company's overall governance and performance in the context of these non-financial factors is of central importance to institutional investors and the ultimate beneficiaries for whom they act, as well as employees, governments and society as a whole. Information is the lifeblood of capital markets. If the information that the market participants have to rely upon is short-term and thin, then these characteristics will define our market.

Unfortunately, the current reporting model, framed by International Financial Reporting Standards, national standards and stock exchange rules, does not provide the necessary framework to enable non-financial factors to be systematically taken into account in reporting and decision-making. Bloomberg data shows that of 25,000 companies surveyed, 75 percent do not report on even one data point of sustainability information (Corporate Knights Capital Study, 2013).

As a result, undue focus and reliance is placed on short-term financial performance, with the risk that capital is not being directed efficiently towards those companies that have robust business models, that make a meaningful contribution towards the achievement of a sustainable society, and which outperform in environmental, social and governance terms.

In order to promote change in this area, in 2008 Aviva Investors called for a debate with stock market listing authorities regarding how they could integrate sustainability issues into the listing rules of stock exchanges. We wanted to explore how exchanges can work together with investors, regulators and companies to enhance corporate transparency, and ultimately performance, on environmental, social and corporate governance issues and encourage responsible long-term approaches to investment.

This initiative has evolved to become the UN Sustainable Stock Exchanges Initiative which is led by the UN Conference on Trade and Development (UNCTAD), the UN Global Compact (UNGC), the UN Environment Programme (UNEP) and the UN-supported Principles for Responsible Investment (PRI). It was named by *Forbes* magazine as one of the "World's Best Sustainability Ideas" in 2010.

While a number of important debates on stock exchanges and sustainability have taken place since this time, these debates have largely focused on encouraging voluntary action by national stock exchanges. A number of exchanges have taken some action – particularly in emerging markets – and this should be welcomed. However, none of the debates so far have managed to reach out beyond the stock exchanges themselves and successfully engage their regulators. This is a key strategic problem as many of the largest exchanges have said they are concerned that they will lose corporate listing business to other exchanges if they move ahead of national regulators and the international community by embedding sustainability disclosure requirements within their exchanges. Conversely, one of the key constraints referenced by capital market participants is the lack of corporate disclosure in the area of environmental, social and corporate governance performance. Consequently, there is a case for internationally coordinated action.

Recognizing this problem, the UN Sustainable Stock Exchanges Initiative produced "A Report on Progress" for their 2012 Global Dialogue. This report included a survey of stock exchanges and found that a minimum level of comparability across markets is needed. At the time, more than three-quarters of stock exchange respondents to the survey welcomed a global approach to consistent and material corporate sustainability reporting. Further, 75 percent of stock exchange respondents favoured internationally coordinated action via a convention on corporate sustainability reporting. The report's recommendations included that "Regulators should work with policymakers in developing an international policy framework requiring, on a comply or explain basis, listed companies to provide material and consistent ESG disclosures", and that policymakers should "set a roadmap for the development of an international policy framework that supports improved and consistent ESG disclosure by listed companies across markets."

Fortunately, therefore, it would appear that the SDGs are likely to address one of the key reasons for market inefficiency from the sustainability perspective. However, our concern is that corporate reporting in isolation is not enough. In order to ensure that the capital markets use and reinforce the information published within the integrated report, the targets need to address the entire system.

Therefore, with reference to Figure 17.1, we suggest targets addressing the entire supply chain of capital. The key point is that by addressing goals that are bespoke to each of the key intermediaries in the capital market chain, the Sustainable Development Goals (SDGs) can help to ensure that the capital markets support a Nexus approach.

The causes of market inefficiencies: A lack of education among market participants on the costs and benefits of corporate sustainability

This is closely related to the corporate sustainability reporting issue where market participants need to be able to see corporate performance. However, simply providing this information does not guarantee that investors will be interested, nor does it ensure that they will know how to use it. This is due to an overall lack of concern among market participants about the costs and benefits of corporate sustainability, which emerges partially due to a lack of knowledge about how much these issues matter.

One practical way of changing this over time would be for the most highly-regarded fund manager and analyst training centres around the world, such as the Chartered Financial Analyst Institute (CFAI), to ensure that their training syllabus and – crucially – the charter holder examination, a test that shapes what is important in the syllabi, improves the ability of analysts to think through how the adoption of Nexus principles will enhance corporate valuation above companies who don't adopt. Universities should also update the content of their finance and business qualifications – particularly the MBA and Masters in Finance – to include sustainability awareness.

The capital markets' *ownership influence* over corporate sustainable development

The capital market's ownership influence over corporate sustainable development and Nexus principle adoption originates mainly from the ownership rights associated with equities. It also exists to a lesser degree with holders of corporate debt. However, debt holders do not have the legal ownership rights that are associated with equities (partly reflecting the lower risks that corporate debt investors take due to them being repaid ahead of equity owners if the company is taken into receivership).

This ownership influence is closely related to the previous discussion on financial influence as investors are expected to voice their concerns to companies about issues that are financially relevant to the company and their holding. However, this is a distinct area as "voice" and "exit" represent alternative courses of action for investors.

So, from the perspective of sustainable development policy makers, what needs to change?

As we noted earlier, the key problem with adopting a Nexus-like approach within the existing capital markets is that the cost of capital for companies is not significantly influenced by the sustainability of the company. This means that financial institutions operating in their short-term self-interest do not need to concern themselves with engaging with companies on sustainable development issues – unless, that is, they are demonstrably relevant to a company and have a potentially material impact on its cash flows. While a great deal of good work has been done by some in this area, engaging with companies on Nexus development issues, particularly in the interlinkages between water, food and energy, remains uncommon. Where it does take place, it can be extremely effective.

However, there are fundamental problems with the ownership chain of influence. For example, it remains extremely difficult for the end investor to know how their AGM votes have been cast, what ownership influence has been exerted, or even, in many cases, where their money is invested. Arguably most reasonable end investors are interested in Nexus issues and do not wish to invest in a way that undermines sustainable development. This opens up possibilities for pushing the Nexus approach to investors.

In economic terms, responsible ownership is a non-excludable public good. That is to say that the benefits of engagement are enjoyed by all owners, regardless of whether they behave as responsible long-term owners by investing in stewardship. Consequently, the vast majority of profit-maximizing commercial fund management institutions free-ride, and either do no real stewardship at all, or invest only token resources in this work.

The question remains, however: what can policy makers do to stimulate this market?

The starting point for policy makers should be to increase the market demand for stewardship in general, and corporate sustainability in particular. It is for asset owners to voice this demand. The best way for asset owners to do this is for them to make it clear in their Investment Management Agreement precisely what is expected when appointing their asset managers. Policy makers can help by providing model contracts that do this well.

Policy makers could stimulate the market further by establishing mechanisms that promote, encourage and require investors to maintain an appropriate oversight role of companies on environmental, social and corporate governance issues.

Our financial fitness tests for the implementation of financing for development and the SDGs

Having set out how capital markets function, and how they relate to the Nexus, we now have a framework to assess the strength of the outcomes on finance within the outcomes from the Financing for Development (FfD) and SDG process is fit for purpose. At the time of writing, the final outcomes from these processes have still to be agreed.

While it is good to see that market mechanisms were being considered, this is still treating symptoms and not cause. Ultimately, the key FfD fit-for-purpose test should be whether the suggested measures will help to ensure that the international capital markets do not finance activities that undermine achievements towards the sustainable development goals. In this area there are some good ideas emerging, but far more needs to be done to put the markets onto a more sustainable footing.

The inclusion of sustainable economic growth and climate change are vital to addressing the financial architecture. The lack of sufficient specificity in particular makes them not sufficiently clear to be actionable. Ultimately, to be considered a success, the implementation process will be judged on these following six key tests:

Test 1. Getting Prices Right: does the outcome recognize the central importance of ensuring that the price mechanism promotes sustainable development?

Test 2. Getting Incentives Right: are there measures that will change the business models and personal incentives of the institutional participants in the capital supply chain – in particular, sell-side brokers, stock exchanges, fund managers, investment consultants and asset owners?

Test 3. Securing Capital: will the changes catalyze investment instruments to finance a Nexus approach that will be sufficiently attractive to markets, and does it look likely to generate a plausible capital-raising plan for these instruments?

Test 4. System-Wide Transparency: does the means of implementation include measures that will promote the transparency of companies on their sustainability performance as well as all the transparency of all the investment intermediaries that connect the end investor to the companies that they own?

Test 5. Sustainable Finance Standards: will the means of implementation create the right kind of hard and soft standards that facilitate sustainable capital markets?

Test 6. Sustainable Demand for Sustainable Finance: does the debate ensure that there is sufficient demand for sustainable finance and sufficient accountability of financial intermediaries by promoting financial literacy measures among the investing public?

Some limitations of this approach

Capital markets are large and complex, so it is important to be clear about the limitations within our proposals. The first and most important limitation is that our proposals do not replace the need for government actions that internalize the many externalities surrounding sustainable development. For example, we have not developed specific proposals on such externalities for companies in a range of sectors, and there are limits to the extent to which reform of the capital markets will be sufficient. Ensuring that the price mechanism works properly is primarily the role of governments, not investors. For example, properly valued environmental and social goods and services lies in government hands. If the economy is to be moved to a truly sustainable basis, then we would expect to see governments taking action to correct the many distortions in the pricing systems on fisheries, freshwater, climate change and natural resource depletion. This is how sustainability issues become relevant to our corporate valuation work, and this is how our capital is put to work in the right places. This requires setting standards, creating fiscal measures such as carbon taxes, or setting up market mechanisms such as carbon trading schemes that price the externalities and ensure that the negative externalities are corrected. In this manner, the Nexus approach can be of vital importance in simplifying and grouping important factors.

In terms of the scope of our recommendations, while Aviva is one of the world's largest insurance companies, the focus of this chapter is on how to promote more sustainable and responsible capital markets. We have chosen to focus in this way as we recognize that the capital collectively raised by the insurance industry can be deployed by asset management companies in ways that will exacerbate some of the hazards that we insure – climate change is an important contemporary example of this. As members of UN Principles for Sustainable Insurance, we play an active role in that debate.

Conclusion

While capital markets are central to the achievement of sustainable development and a Nexus approach, they currently do not need to understand or reward sustainable behaviour. This is either because the markets are inefficient and do not reward good behaviour, or because market failures mean that investors do not need to worry about the very long-term costs because they are outside of their investment time horizons.

There are measures that we can take now to change this problem. This is particularly true where the issue is simply that markets are inefficient and do not reward good behaviour. For example, the transformation to a sustainable economy should focus on the incentives of all key players within the capital market such that they are sanctioned and incentivized on their sustainability performance. We also need much better market information, which will almost certainly require a change to global listing rules that mandate the disclosure of strategic sustainability reports, and provide the owners of companies with an opportunity to vote at the company's AGM on the report. Better training of market participants on the materiality of sustainability issues as well as how they can be factored into valuation analysis would also help.

However, before capital markets can be genuinely sustainable, we need capital market policy makers to have greater regard for future generations when setting policy. For the health of the economy, society and the environment, policy makers should integrate sustainable development issues into capital market policy making, and the Nexus approach provides a good way to bundle these issues. We need policy makers to internalize corporate externalities onto company accounts via, for example, the increased use of fiscal measures, standards and market mechanisms. We also need to ensure that the culture within global financial services firms is not one where the many conflicts of interest are exploited. Some of this will require greater government intervention, particularly around the regulation of investor action on responsible ownership.

The SDGs and the outcome of the Financing for Development process need to pass six tests: (i) getting prices right; (ii) getting incentives right; (iii) securing capital; (iv) securing system-wide transparency; (v) catalysing sustainable finance standards; and crucially, (vi) catalysing sustainable demand for sustainable finance.

In this way, capital markets will become the primary facilitator of a green and just global economy that helps to resolve the Nexus of issues surrounding poverty, climate change, fresh water and human rights.

References

Corporate Knights Capital (2013) *Trends in Sustainability Disclosure: Benchmarking the World's Stock Exchanges*, Toronto: CKC. Available online at: http://www.corporateknights.com/reports/2013-world-stock-exchanges/

Eccles, G.E., Ioannou, I. and Serafeim, G. (2013) *The Impact of Corporate Sustainability on Organizational Process and Performance*, Working Paper, Boston: Harvard Business School. Available online at: http://hbswk.hbs.edu/item/the-impact-of-corporate-sustainability-on-organizational-process-and-performance

NAPF/IMA (2004) *NAPF/IMA Short-Termism Study Report*, London: MORI.

Stigson, B. (2003) *Speech: Walking the Talk – The business case for sustainable development*, Geneva: WBCSD.

UNEP (2010) *Translating ESG into Sustainable Business Value*, Geneva: UNEP. Available online at: http://www.unepfi.org/fileadmin/documents/translatingESG.pdf

18 Principles for the integration of the Nexus within business

Felix Dodds and Cole Simons

Introduction

In the context of this chapter, the Nexus refers to four areas: water, food, energy and the impacts of climate change. We are suggesting a set of Nexus Principles for companies, designed to draw together previous principles set forth for individual sectors. These include the United Nations Global Compact (UNGC) Food and Agriculture Business Principles, the UNGC CEO Water Mandate, the Policy Recommendations from the Bonn 2011 Water, Energy, Food Security Nexus Conference and the Principles from the Nexus Water, Food, Climate and Energy 2014 Conference Declaration.

Why the Nexus?

In 2011 the Stockholm Environment Institute (SEI) paper 'Understanding the Nexus' suggested that by 2030 there will be an increase in global population by one billion, that we will experience an increasingly urbanized population that comprises more than 60 percent of the total, and that changing economic activities of fast developing countries, such as India and China, will bring additional stresses on the world. The predictions they made also included that there will be a shortfall of water over the next fifteen years with as high as 30 to 40 percent demand above availability, as well as an increase in demand for food and energy of around 40 percent. All of this is made worse by climate change and the lack of an effective agreement to keep global temperature increase at under 2 degrees by 2100.

Such an agreement would require trade-offs between the different sectors of water, energy and food. Traditionally these sectors have operated on an individual basis, in particular competing for water in an increasingly water-scarce world. There has been some excellent work undertaken by companies such as SABMiller, Pepsi, Coke, H&M and Levi's to reduce their water footprint, and by companies like AVIVA and Bloomberg in raising concerns within the financial markets. The push to require all companies listed on stock exchanges to report and explain their environmental impact, governance and social policies and practices would go a long way to encourage companies to adopt a Nexus approach. Two reports to the UN Secretary-General, delivered in 2011 and 2013, by high level, expert panels, suggested that this should be done. Steve Waygood of AVIVA explains this in more detail in Chapter 17. By adopting the principles below and adopting a reporting process, companies can use the Nexus to ensure a more sustainable future not only for their own companies but for all of us.

The Nexus concept acknowledges that at the intersection of the three sectors of water, food and energy the impact of climate change will detrimentally affect each, and that an integrated approach is required to minimize externalities, thus reaching an optimal solution. To create optimal solutions it is necessary to better understand the interlinkages of company strategies, investment and behaviours. Specifically, this chapter will suggest a set of six Nexus Principles to guide the development of company strategies as they further develop and analyze the connections between water, food, energy and climate change in order to promote local, national, transboundary and global solutions now and in the future.

Principle 1: Respect human rights to ensure all benefit from the Nexus

Businesses should provide a healthy work environment that protects and furthers human rights. Business should respect human rights within all the communities where they operate. This is a critical foundation for Nexus solutions to be possible.

In developing this we have drawn from:

- The UNGC's Food and Agriculture Business Principles, which suggest a very similar principle:

> Respect Human Rights, create decent work and help communities thrive. Businesses should respect the rights of farmers, workers and consumers. They should improve livelihoods, promote and provide equal opportunities, so communities are attractive to live, work and invest in.
>
> (UNGC, 2014)

- The UNGC's CEO Water Mandate does not possess a core element that deals with respecting human rights. However, in its revised Preamble it does say:

> We note the 2010 resolutions by the UN Human Rights Council and the UN General Assembly recognizing the human right to safe drinking water and sanitation.
>
> (UNGC, 2007)

- It recognizes that collective action and community engagement can be construed to be a testament to respecting human rights without saying it expressly:

> While individual organizational efforts will be critical in helping to address the water challenge, collective efforts – across sectors and societal spheres – will also be required. Such multi-stakeholder collaboration can draw on significant expertise, capacities and resources. Utilizing frameworks such as the UN Global Compact, companies can participate in collective efforts to address water sustainability.
>
> (UNGC, 2007)

- Bonn 2011 Policy Recommendations 1: Nexus challenge: This says that decision-making starts with:

> Central to a human rights approach is the achievement of water, energy and food security for the poorest of the poor.
>
> (Bonn, 2011)

- The 2014 Nexus Declaration underscores that:

> Long-term solutions must be grounded in a human rights framework with an explicit recognition of the rights of women, including a fulfilment of the unmet demand for voluntary family planning to slow rapid and unsustainable population growth.
>
> (Water Institute, 2014)

Such ethical actions remain pertinent because there is no way businesses can grow a sustainable framework if they do not operate responsibly and in tune with the needs of the communities surrounding them. Recognizing human rights is a fundamental requirement in this.

Principle 2: Ensure integrity, transparency and accountability in addressing the Nexus

Businesses should act on core values of integrity, transparency, and accountability (avoiding corruption) such that they build inter-community trust and collaboration so that Nexus decisions are a common goal.

In developing this we have drawn from:

- UNGC's Food and Agriculture Business Principles: this is not included as an individual mandate or principle, but this principle is alluded to across the entire document. In the governance principle they say:

> Businesses should behave legally and responsibly by respecting land and natural resource rights, avoiding corruption, being transparent about activities and recognizing their impacts.

This is largely similar to the Nexus principle 2.

- CEO Water Mandate: the CEO Water Mandate possesses a principle regarding transparency, a small part of the Nexus principle 2:

> Transparency goes to the heart of accountability. Leading companies recognize that transparency and disclosure are crucial in terms of meeting the expectations of a wide group of stakeholders. Such efforts help companies focus on continuous improvement and turning principles into results – a process which is crucial in terms of realizing gains and building trust.
>
> (UNGC, 2007)

- Bonn 2011 Policy Recommendations 3: Opportunities to make a difference underscores this:

> Promote access, productivity gains and more equitable sharing of benefits through explicit commitments to transparency and integrity systems. Corruption takes many forms and can be widespread in major infrastructure projects and management systems resulting in economic and financial losses and distortions in decision-making, including in relation to utilization of resources. High level support for efforts to respect ethical behavior, reduce corruption and enhance integrity, transparency and accountability needs to be followed through the system at all levels, and across and within sectors, with awareness raising, the adoption of available good practice tools and more effective enforcement.
>
> (Bonn, 2011)

By fostering a transparent system, businesses enable input from the communities within which they operate and ensure that these communities understand the decisions and decision-making processes companies have undertaken when creating the most mutually beneficial system with the least amounts of externalities and negative impact. This is exactly what the Nexus approach is trying to create.

Principle 3: Support government attempts to build a sustainable Nexus regulatory framework

Businesses should support governments at all levels in the creation of regulatory frameworks through consultation with relevant stakeholders to ensure that the best trade-offs are made possible.

In developing this we have drawn from:

- UNGC Food and Agriculture Business Principles: including a principle entitled: "Encourage good governance and accountability." This states that:

> Businesses should behave legally and responsibly by respecting land and natural resource rights, avoiding corruption, being transparent about activities and recognizing their impacts.
>
> (UNGC, 2014)

This is an approach based on compliance; it does not recognize businesses, potential to help promote positive governmental interventions.

- UNGC CEO Water Mandate: Principle 4 within the CEO Water Mandate illustrates much of which the Nexus principles hopes to integrate. In Principle 4, the CEO Water Mandate states that companies should:

> Contribute inputs and recommendations in the formulation of government regulation and in the creation of market mechanisms in ways that drive the water sustainability agenda. Endorsers should seek ways to facilitate the development of good policy.

By doing so it clearly sees a more proactive role in the development of new regulation and public policy.

- Bonn 2011 Policy Recommendations 4: Taking action: scope, roles and responsibilities focuses more on the role of local authorities and utilities:

> With adequate regulation and the right incentives, the inherent innovation within business and research centers can provide solutions and attract finance, particularly in planning for rapidly emerging urban centers that do not yet have the congestion and complexity of established megacities.

> It fails to deal with the role of regulation in creating an enabling environment for trade-offs neither within the Nexus nor for a positive role for companies in developing the regulation.

> Active governmental intervention at all levels is required to develop an effective enabling environment for trade-offs within the nexus and to achieve this there will need to be the involvement of all stakeholders, including companies. The nexus principles recognize that companies should be supporting efforts to create good legislation. Their actions are not just positive for them as a company and their immediate benefactors, but also for the community as a whole. In this way the company creates a culture in which the company is seen by the community and the government as a positive partner.
>
> (Bonn, 2011)

Principle 4: Promote co-design and mutually beneficial partnerships

Businesses should work on the creation of mutually beneficial partnerships with relevant stakeholders and governments that work to enhance the spread of positive action on the Nexus and increase optimal returns for businesses.

In developing this we have drawn from:

- Food and Agricultural Business Principles: these have two principles which can be summed up in Nexus Principle 4. The two principles so named are "Ensure economic viability and share value" and "Promote access and transfer of knowledge, skills and technology."
  Nexus Principle 4 seeks to combine these two to promote the need for the community/stakeholders to work together in a continual conversation about the most sustainable plan and the best methods to achieve that while working with others.
- UNGC CEO Water Mandate: Principle 3 of the CEO Water Mandate on 'Collective Action' is similar to Nexus Principle 4. The CEO Water Mandate recognizes the need for all organizations and people to work together in reaching dependable and sustainable solutions:

> While individual organizational efforts will be critical in helping to address the water challenge, collective efforts – across sectors and societal spheres – will also be required. Such multi-stakeholder collaboration can draw on significant expertise, capacities and resources. Utilizing frameworks such as the UN Global Compact, companies can participate in collective efforts to address water sustainability.
>
> (UNGC, 2007)

- Bonn 2011 Policy Recommendation 3: Opportunities to make a difference goes on to say:

> Cooperation is needed at the national level through strategic planning and functional linkages to coordinate sectoral ministries and other stakeholders including civil society; in business through strengthened incentives, public–private partnerships and improved corporate responsibility programs.
>
> (Bonn, 2011)

- The 2014 Nexus Declaration addresses this when it says:

> New institutional arrangements that move beyond sectoral or silo governance structures can encourage cooperation between governmental departments and form partnerships with business and other stakeholders that discourage silo thinking, tackle cross-sectoral problems, and implement integrative solutions.
>
> (Water Institute, 2014)

Without an open dialogue about how best to tackle unsustainable actions and trade-offs, solid options for minimizing negative impacts will be hard to configure and improve upon. Likewise, by not sharing possible alternatives, no ideas will be adequately tested and analyzed. This could lead to systems that, although created with the best intentions, may not be as effective as they would be if they had community and peer interaction and feedback. Partnerships between companies, other stakeholders and governments at the relevant level will ensure a more effective approach to addressing the Nexus.

Principle 5: Seek innovative methods of sustainable resource usage

Businesses should strive to find and implement resource use in a manner that further mitigates environmental impact and reduce waste.

In developing this we have drawn from:

- UNGC Food and Agriculture Business Principles: approach this through the call to 'Be Environmentally Responsible'. The Principles call for businesses to:

> Support sustainable intensification of food systems to meet global needs by managing agriculture, livestock, fisheries and forestry responsibly. They should protect and enhance the environment and use natural resources efficiently and optimally.
>
> (UNGC, 2014)

- UNGC CEO Water Mandate: Principle 1 on 'Direct Operations' is the closest those principles come to addressing Nexus Principle 5 as well as the Nexus Principle 6. The Water Mandate Principle 1 states that companies should:

> Set targets for our operations related to water conservation and waste-water treatment, framed in a corporate cleaner production and consumption strategy. Endorsers should reduce water waste discharge without providing any focus on the need to alter their approach in managing the situation.
>
> (UNGC, 2007)

- The 2014 Nexus Declaration addresses this when it says:

> Unsustainable patterns of production and consumption are leading to severe pressure on planetary boundaries that must be reduced to avoid further damage to planetary systems and ensure critical tipping points are not reached.
>
> (Water Institute, 2014)

- Bonn 2011 Policy Recommendation 3 Opportunities to make a difference, in its section entitled 'create more with less', goes on to talk about the need to increase the efficiency of "resource use in manufacturing and production processes and publicize innovations and good practice for demand management, increasing efficiency and raising productivity" (Bonn, 2011).

This principle recognizes the fact that there are issues in the planning of individual sectors and forward action that must be taken together in order to solve these issues in a more productive way in the future. In this principle and Principle 6, the Nexus principles look to set forth a call to action where companies actively look to create systems of change that not only benefit the environment around them but also place them ahead of the competition. Specifically, Principle 5 recognizes that we live in an increasingly resource-scarce world.

Principle 6: Incorporate Nexus views in future development

Businesses should consider these Nexus principles in the development of future business plans and goals in an effort to alleviate externalities. Businesses should also provide the ability to discover and integrate technologies and policies that further sustainability in future plans.

In developing this we have drawn from:

- UNGC Food and Agriculture Business Principles: The closest to this is 'Aim for Food Security, Health and Nutrition'. This principle expects that businesses should:

> Support food and agriculture systems that optimize production and minimize wastage, to provide nutrition and promote health for every person on the planet.
>
> (UNGC, 2014)

- UNGC CEO Water Mandate: Once again, Principle 1 goes 'beyond conventional decision making' where it refers to:

> Mutually beneficial responses and provides an informed and transparent framework for determining trade-offs to meet demand without compromising sustainability and exceeding environmental tipping points. It aims to bring economic benefits through more efficient utilization of resources, productivity gains and reduced waste. This is a particular challenge in today's economic climate, yet the consequences of inaction would become increasingly severe on people's welfare, economic growth, jobs, and the environment.
>
> (UNGC, 2007)

Bonn 2011 Policy Recommendation 3: Opportunities to make a difference addresses this principle in part through the section on ending waste and minimizing losses when it says:

> Life cycle analysis will ensure externalities are incorporated and any perverse subsidies that favor natural resource exploitation over re-use are recognized and can be debated. In the case of energy production, multiple benefits may be achieved by promoting second-generation technology bioenergy but need to be viewed in the context of the agricultural system as a whole. Avoiding redundancy in product design and extending serviceability can also make a contribution to reducing waste and lowering demand.
>
> (Bonn, 2011)

- The 2014 Nexus Declaration address this when it says:

> The world is a single complex system in which all the parts and subsystems constantly interact. Global problems such as persistent poverty and climate change should be viewed from this perspective and solutions and policy interventions should be sought that are beneficial for the system as a whole.
>
> (Water Institute, 2014)

Principle 6 largely builds on the call to action of Principle 5. Where Principle 5 looks more towards the current state of issues and finding the best option for now, Principle 6 looks to create an environment of planning within a business that anticipates the need to develop a sustainable method for continually improving its actions in the future. This principle also serves to tie in previous principles by recognizing the need to integrate human rights, act in a transparent manner, and engage communities and other forward-looking companies in a mutually beneficial manner.

Embedded in the complex relationships between water, energy, food and the impacts of climate change is societal well-being, which has many attributes we do not yet understand. Business, NGOs, academic and other stakeholders and governments at all levels will need to build relationships that are flexible and work in the future, but that maintain both collaboration and humility. Collaboration will allow us to appropriately apportion the risks and benefits associated with moving forward with appropriate solutions. Humility will allow us to step back, examine outcomes and returns, and adjust our thinking together so that solutions of the future are better than those that we might imagine today.

Conclusion

We are starting to recognize that we live in a complex world and that we need to address issues such as water, energy, food and climate change together. In the past we dealt with these issues in silos. If we consider the issues from a Nexus approach then we largely eliminate contradictory acts and make outputs vastly less complicated. Before, we attempted to account for individual sectors with individual programmes and principles. By recognizing that these sectors are interconnected, we have presented principles that if followed will allow for more effective management of systems, with businesses crafting policies to recognize this interdependency and thus simplify the approach.

In effect, this set of principles ties together the needs and calls to action of the documents we have used as references in a simple form, allowing more immediate and successful action, just as the Nexus approach pulls together multiple other sectors for more prudent and productive decision making.

References

Bonn Nexus Conference (2011) *Nexus Policy Recommendations*, Bonn: German Government. Available online at: file:///C:/Users/Felix/Downloads/bonn2011_policyrecommendations%20(2).pdf

Stockholm Environment Institute (2011) *Understanding the Nexus*, Stockholm: Stockholm Environment Institute. Available online at: http://www.water-energy-food.org/en/whats_the_nexus/background.html

United Nations Global Compact (2007) *CEO Water Mandate*, New York: UNGC. Available online at: http://ceowatermandate.org/files/Ceo_water_mandate.pdf

United Nations Global Compact (2014) *Food and Agriculture Business Principles*, New York: UNGC. Available online at: https://www.unglobalcompact.org/docs/issues_doc/agriculture_and_food/FABPs_Flyer.pdf

Water Institute (2014) *Building Integrated Approaches into the Sustainable Development Goals: Nexus Declaration*, Chapel Hill: Water Institute. Available online at: http://nexusconference.web.unc.edu/files/2014/08/nexus-declaration.pdf

Index